"You're Muted"

"You're Muted"

Performance, Precarity, and the Logic of Zoom

Edited by
Mark Nunes and Cassandra Ozog

BLOOMSBURY ACADEMIC
NEW YORK • LONDON • OXFORD • NEW DELHI • SYDNEY

BLOOMSBURY ACADEMIC
Bloomsbury Publishing Inc, 1359 Broadway, New York, NY 10018, USA
Bloomsbury Publishing Plc, 50 Bedford Square, London, WC1B 3DP, UK
Bloomsbury Publishing Ireland, 29 Earlsfort Terrace, Dublin 2, D02 AY28, Ireland

BLOOMSBURY, BLOOMSBURY ACADEMIC and the Diana logo are trademarks of
Bloomsbury Publishing Plc

First published in the United States of America 2024
Paperback edition published 2026

Cover design: Eleanor Rose
Cover photograph: Paula Krasiun-Winsel/Tidepool Productions

"You're Muted": Performance, Precarity, and the Logic of Zoom examines the rapid
emergence of videoconferencing in everyday life under Covid-19, its preexisting
performative logic, and the ongoing implication of these practices, through the frame of
Zoom. Zoom is a registered trademark of Zoom Video Communications, Inc. or other group
companies ("Zoom"). This book is in no way connected to, affiliated with, or endorsed by
Zoom or the Zoom brand.

Portions of the book introduction and section introductions appear in an early version as
Mark Nunes and Cassandra Ozog, "Your (Internet) Connection Is Unstable," *M/C Journal*
24, no. 3 (2021). https://doi.org/10.5204/mcj.2813.

An earlier, condensed version of "Mediating the Death of a Parent: Zoom and the Deathbed
Vigil" appeared as Susan Sci, "By Mom, I Love You," *In Media Res*, August 24, 2021. https://
mediacommons.org/imr/content/bye-mom-i-love-you.

A catalog record for this book is available from the Library of Congress.

ISBN: HB: 979-8-765-10824-6
PB: 979-8-7651-0825-3
ePDF: 979-8-765-10827-7
eBook: 979-8-765-10828-4

Typeset by Deanta Global Publishing Services, Chennai, India

For product safety related questions contact productsafety@bloomsbury.com.

To find out more about our authors and books visit www .bloomsbury .com and sign up
for our newsletters.

To Matthew, Joshua, and Hayleigh, who experienced the pandemic as rite of passage, bardo, *and trial by fire.*
To Eric and Penny, who inspire me to keep fighting for a better world.
And to all the students, academics, writers, and artists who struggled through the past few years to live, create, and survive on Zoom. We see you.

Contents

Illustrations

Figures

Table

Acknowledgments

We would like to thank the following individuals for their help in making this collection of essays a reality. First off, we recognize the New Media and Digital Cultures Working Group of the Cultural Studies Association, which came to life as a working body in the midst of the pandemic, by way of Zoom. Were it not for that forum, the coeditors would not have had the opportunity to collaborate on this project. We would also like to thank Axel Bruns and *M/C Journal*, where we first explored some of these concepts. We would also like to thank Paula Krasiun-Winsel/Tidepool Productions for technical assistance on the images included in this text and for her creative and thoughtful work in capturing our cover image. It has been a pleasure to work with Katie Gallof and Stephanie Grace-Petinos and their entire editorial team at Bloomsbury, who have helped us work our way through matters both mundane and arcane. We would also like to acknowledge the support of our departments and institutions—the Department of Interdisciplinary Studies at Appalachian State University and the Department of Sociology and Social Studies at the University of Regina. And, finally, we would like to thank our partners and family for their patience and support during long hours, late nights, and excessive Zoom use.

Introduction

The Performative Logic of Zoom

Mark Nunes, Appalachian State University, and Cassandra Ozog,
University of Regina

As 2023 draws to a close, mass media and social media alike have grown weary of Covid-19, contributing to the popular sentiment that the pandemic is finally "over." We are, after all, more than three years past that tumultuous moment when the World Health Organization (WHO) declared the Covid-19 outbreak a global pandemic and six months beyond the end of the pandemic as a "public health emergency of international concern" (World Health Organization 2020, 2023). Yet to declare ourselves "post-pandemic" is highly problematic not only from a public health standpoint[1] but also because many of the organizational and institutional restructurings that occurred in the rapid response to an international health emergency have remained in place. Videoconferencing as a feature of everyday life for millions marks one such site: a *sigil* of the dramatic social and cultural transformation that occurred during "pandemic times" but likewise an index of all that has emerged as the "new normal" since March 11, 2020. The uncanny familiarity of these practices that are now embedded in a wide range of everyday uses—from remote work to hybrid classrooms to telehealth sessions—likewise signals the degree to which the technological and ideological apparatuses supporting this new normal were long in place well before the first stay-at-home order. This collection of essays examines those conditions, and their ongoing implications, through the all-too-familiar frame of Zoom.

The Performative Logic of Zoom

The ubiquity of Zoom as a videoconferencing platform makes it almost unnecessary to introduce the brand. Founded nearly a decade before the pandemic, Zoom had already established a significant footprint in both business and educational settings by the time it went public in 2019.[2] By the end of its first fiscal year as a publicly traded company––when confirmed Covid-19 cases in the United States still numbered in the single digits––Zoom Video Communications, Inc. (2020a) was reporting "approximately 81,900 customers with more than 10 employees" and a total yearly revenue of over $622 million. As cases grew logarithmically and spread globally in early 2020, and with the possibility of lockdowns and stay-at-home orders seeming imminent, Zoom's adoption rates soared. By April 30, 2020, just six weeks after the WHO pandemic declaration, Zoom was reporting 265,400 customers with more than ten employees––more than triple what it reported three months earlier––and a single quarter revenue of over $328 million (Zoom Video Communications 2020b). Those numbers continued to grow throughout 2020 and onward, as did Zoom's support of free use of the platform for personal as well as K-12 education.

Overnight, it seemed, Zoom became the default platform for videoconferencing, rapidly morphing from brand name to eponymous generic—a verb and a place and mode of being all at once. It is for this reason, in part, that this volume focuses predominantly on Zoom, although many of the insights and observations shared on these pages will apply as well to other personal and enterprise-grade videoconferencing platforms—be that Microsoft Teams, Google Meet, Skype, or FaceTime. We are interested less in the company Zoom Video Communications, Inc. than in *the apparatus of Zoom* and the logic it expresses: what Crano (2021) refers to as Zoom's *dispositif* of framed faces in networked encounters and "its promise [of] sustain[ed] social harmony and economic efficiency through an extended public health crisis and waves of state-mandated isolation" (see also Crano, this volume). The ease at which the "genericization" of Zoom occurred in the midst of such widespread social upheaval calls attention to the fact that the logic through which videoconferencing operates predates, predicates, and to some degree *predicts* that crisis. Thus "Zooming" in its generic expression of video-mediated interaction presented itself as both a novel mode of engagement during times of social distancing and a rather banal modification in networked presencing that seemed hardly new at all. As Lovink (2022) notes, "within weeks video calling moved from a global experiment to a foregone conclusion" (33).

Without question, something profound shifted in the practices of everyday life for tens of millions of individuals as they "migrated" and "pivoted" from in-person work settings to videoconferencing platforms, and millions more turned to mediated proximity to connect with family, friends, and loved ones during a time of restricted travel and limited in-person contact. But while this nearly ubiquitous transition to remote work and remote play was unprecedented, it was also entirely anticipated to the extent that teleworking, digital commerce, online learning, and social networking were, by 2020, rather common fare. For millions, of course, lockdowns and travel restrictions had a profound personal, social, and economic impact that could not be mitigated by the mediated presence offered by way of Zoom and other videoconferencing platforms. For those of us privileged enough to have modes of labor that allowed us to engage in work remotely, however, Zoom tapped into a network logic that was already governing our neoliberal work lives long before the first lockdowns began.

We trace this logic back to what Lyotard (1984) identifies in *The Postmodern Condition* as a "logic of maximum performance" that regulates the contemporary moment: a cybernetic framework for understanding what it means to *communicate*—one that ultimately frames all political, social, and personal interactions within matrices of power laid out in terms of performativity and optimization (xxiv). We likewise connect this cybernetic frame, and its emphasis on control and the regulation of flow, to earlier incarnations of network logic dating back to the development of postal roads and daily newspapers routes (Nunes 2006) but accelerating considerably under the "control revolution" (Beniger 1986) developments in industry and commerce in the nineteenth century. Under this logic, "communication" becomes synonymous with *transmission* and the technological and ideological apparatuses that support "the transmission of signals and messages over distances for the purpose of control" (Carey 1992, 15). For Lyotard, "performativity" as it presents itself in Europe and the United States during the post-Second World War period presents a further acceleration of that logic, driven by a systems theory view of social and economic relations. He writes: "The goal of this system, the reason it programs itself like a computer, is the optimization of the global relationship between input and output–in other words, performativity" (11). This logic of performativity serves as a foundation for not only how a system operates but for how all other elements within that system express themselves. Performative control systems, however, need not exert "total control" to the extent that variation, modulation,[3] and adaptation may, in fact, create circumstances of increased productivity and maximized performance (Lyotard 1984, 55–6). Thus, Lyotard explains:

> Even when its rules are in the process of changing and innovations are occurring, even when its dysfunctions (such as strikes, crises, unemployment, or political revolutions) inspire hope and lead to a belief in an alternative, even then what is

actually taking place is only an internal readjustment, and its results can be no more than an increase in the system's "viability" (11–12).

And one may well add to Lyotard's list of systematic dysfunctions *a global pandemic.*

This framework helps to explain why the massive uptake of Zoom in the midst of a "public health emergency of international concern" simultaneously signaled both a crisis of performativity and its technological salvation. Zoom, in effect, offered universities, corporations, mass media outlets, and other organizations a *platform* (in every sense) to "innovate" within an ongoing network logic of maximum performance: to maintain business as usual in a moment in which nothing was usual, normal, or functional (Grandinetti 2022). While it is easy to point to the rapid increase in remote workers as a measure of systemic, economic disruption, these same numbers suggest how the crisis of a global pandemic only served to affirm the resilience of this performative logic for those workers whose economic and educational advantage had already positioned them in a mode of labor in which "for the most part, the responsibilities of their job can be done from home" (Pew 2020). As Roy (2022) highlights for a *Forbes* piece on the "past, present and future of remote work," the threefold increase in Americans who work from home (WFH) reveals the deep connections between business practices and the technology supporting those practices: performative values framed as gains in productivity, access to a global workforce, and a reduction in office infrastructure costs. Zoom, in its promise of ongoing connectivity and a performative transcendence over social distancing mandates, offered the material and conceptual framework for ongoing institutional operations that was so desperately needed in the face of social and economic disruption—but it could only do so to the extent that these technological and ideological apparatuses were already in place: a *virtual* reality that needed only be actualized when the dysfunction of a global pandemic prompted a modulation in performance.

In our not-quite-post-pandemic moment—after the end of stay-at-home orders and social distancing guidelines—a growing number of businesses are now calling employees back, and these "requests" have increased in volume and intensity during the second half of 2023 (Goldberg 2023b). But this call for bodies to return to cubicles is hardly a sign that "Zooming" is a thing of the past; rather, we identify in this moment a symptom of yet another modulation in a network of maximum performance. The technological solutionism (Morozov 2013) that drove such a rapid and widespread adoption of Zoom and other videoconferencing platforms in the midst of a global health crisis spoke less to remote work as a *solution* to pandemic disruptions than it did the corporate and institutional confidence that those same technological and ideological apparatuses could continue to assert their authority *as* a solution, even in the face of a systemwide, performative disequilibration. As Ricky D. Crano touches upon in the first chapter in this collection, and as we discuss in greater depth in our own chapter, Zoom and other videoconferencing platforms created the illusion of ongoing, optimal performance by casting to the borders and margins of the frame those irruptions of dysfunction that are, in fact, central to the operation of the platform itself. Nothing was *status quo* under the "new normal" marked by precarious encounters and unstable connections. As Goldberg (2023b) notes, the performative adjustments and "innovations" that arose during Covid-19 amounted to a form of an "impromptu experiment" for the corporate sector, one that forced "executives . . . to discern how to make multimillion-dollar decisions in between bursts of 'you're on mute,' [and] employees figuring out how to forge friendships and nudge mentors for advice while sitting next to piles of their laundry." Once again, however, the performativity principle under which corporations and institutions operate casts such dilemmas as outlying aberrations rather than an operational parameter— declaring the feature is really a glitch—and thereby systematically erasing the lived reality that *our connections are always unstable* in a network society.

But as companies (including Zoom) increasingly demand that their employees return to work, the work to which they return remains deeply hybridized (Goldberg 2023a)—a form of labor that requires a set of dispositions that blur the lines between work and home and between corporate persona and private life. As Jacquelyne Thoni Howard notes in her chapter in this collection, these hybrid borderlands create differential structures of precarity for workers, mapped by race, gender, and ethnicity, and do so in ways that correlate with workplace performance metrics. Discussions and debates regarding proper Zoom etiquette for different settings, for example, and what constituted work-appropriate attire when working from home, became particularly gendered in nature (London Image Institute 2021; Howard this volume). Likewise, as Roberts and McCluney (2020) note in a *Harvard Business Review* article written in the Covid-19 summer of 2020, videoconferencing from home for Black Americans limited the ability to code-switch between a racially encoded set of standards for corporate professionalism and a private life at home experienced as a more authentic self. They write: "Scrutiny of Black worker's bodies, cultural aesthetics, and home life are heightened in the virtual workspace, and can exacerbate the already stressful conditions Black workers are experiencing during this pandemic" (see also

Archer et al. 2024). Hybridity suggests a mode of adaptability (most classically, perhaps, in the corporate cliché of *the pivot*, discussed later) in which workers find their role within a logic of maximum performance, one in which the affective burdens that come with intensified surveillance of those who occupy the most precarious positions within these networks of power take a secondary place to the mandate to maintain the flow of the networks themselves. Such is the structure of communicative capitalism (Dean 2009) in the age of Covid-19.

Zoom plays into this performativity logic to the extent that, as its earliest company descriptions declared, its mission is to provide technological solutions that would "bring . . . teams together in a frictionless environment to get more done" (Zoom Video Communications Inc. 2020a). The "frictionless" here certainly has its origins in Bill Gates's (1995) vision, laid out in *The Road Ahead*, of a "friction-free capitalism." Friction, apparently, is a bad thing—to the extent that it speaks to a slowing down, marking those moments (technologically solvable, of course) in which performance fails to meet its maximum potential. Friction, then, also comes to mean those who present a problem to the smooth operations—read: working potential—of Zooming. Folks with unstable internet connections or lack of dependable technology, those who may face language barriers and/or lack of technological knowledge (such as seniors and immigrants, as discussed by Evgeniya Pyatoskaya and Earvin Charles B. Cabalquinto in this volume) thus limiting their frictionless presence on Zoom, or folks who may struggle to balance family and home demands while also trying to WFH, making their "productive hours" less than, well, *productive*. Marx's (1992) concept of labor as a "peculiar commodity," as we are valued only when productive, extended to those who could not use videoconferencing perfectly, regardless of the larger class, race, and gendered demands that played a major role in causing such unwanted frictions. The internet, and information communication technologies (ICTs) in general, promise to fulfill a neoliberal dream of seamless market engagement, be that through frictionless commerce (Lemieux and Duff 2020), frictionless enterprise (Agarwal 2018), or frictionless payment systems (Wood 2023). As with the frozen Zoom windows and muted team members that we treat as aberrant moments rather than critically embedded within the same infrastructure, "frictionless" treats materiality itself—those "friction points" that slow down otherwise immaterial exchanges of capital, content, and data— as latent opportunities to minimize lost productivity and less-than-optimal performance. Well into 2023, Zoom (2023a) still made reference to their service as a "frictionless communications platform" that offered "a space where you can connect to others, share ideas, make plans, and build toward a future limited only by your imagination."[4] Under the logic of Zoom, it is only the imagination that limits "frictionless" productivity, not the bodies that perform work or the social and cultural contexts in which such labor occurs.

Performativity in the "Zoom University" Classroom

As in the corporate space, at colleges and universities (and to a lesser extent, perhaps, primary and secondary schools as well), the "overnight" shift to online instruction in the spring of 2020 speaks less to the "can-do-ism" of the educational system in developed countries than it does to the logic of maximum performance that had already enmeshed itself in the business of education. Much of what has been written about the rise of the neoliberal university over the past three decades calls attention, implicitly or explicitly, to the performativity principles that now dominate educational standards, with the goal of achieving a more efficient, effective, and outcomes-oriented institutional framework for teaching and learning (Roberts 1998; Nunes 2006; Kitto and Higgins 2010; Cardinal et al. 2023). In this regard, what John McArthur refers to in this volume as the "classroom pivot" from face-to-face instruction to Zoom classroom was indeed a profound disruption to pedagogical practices but also a shift that occurred so rapidly as to suggest that the potential for such a redefining of the learning environment had been long in place: real as a potential but yet to be actualized. It is hardly coincidental that academia would embrace "the pivot"—what *CEO Magazine* described as potentially "the business buzzword of the year" for 2020—to describe this sudden shift in teaching modality as an opportunity to "adapt quickly and innovate boldly" in the face of systemic challenges ("Power" 2020). However clichéd the term may be, framing this adaptive response to dysfunction as a "pivot" highlights the same performativity principle at work in the university as in the corporate world, in which innovation in the face of crisis reaffirms a faith in the resilience of an underlying network logic. As Zoom (2022) itself declares in its "The Future of Education" report:

> The events of 2020 and 2021 saw educational networks of all levels, styles and sizes pivot to support teaching and learning in new ways. As educators rushed to try different ways to deliver teaching and learning remotely, and experimented with pedagogy, their institutions scrambled to implement the right technologies and practices to support teachers, students and families. These two painful years

will be remembered both as chaotic and stressful, but also as a Petri dish for educational innovation.

This performativity principle, then, does not simply map a set of performance measures imposed upon students, teachers, and learning institutions (although such measures certainly abound in the neoliberal university). Rather its logic likewise determines the environment in which individuals must orient their teaching and learning performance.

The "Future of Education" report, written in partnership with Australia-based consultant firm Intelligent Business Research Services, goes on to foreground "digital thinking" as key to both "resilience" and "innovation"—transformation that impacts not only instructional delivery but curriculum itself, favoring "highly granular, bite-sized educational content that can be reused across multiple curricula and upgraded frequently" (Zoom Video Communications Inc. 2022, 12). As Pyatovskaya explains in this volume, the forms in which resilience occurs serve likewise as markers of how well integrated individuals and institutions are within the dominant logic that determines access to resources. As she notes, even in stories of success, this resilience "make[s] visible the very inequalities that triggered resilience." The performative logic that Zoom so well articulated during the height of the pandemic need not hide these disparities. In fact, the technological solutionism that it embodies can simultaneously attempt to call out this inequity as an injustice while at the same time exacerbating the circumstances that necessitate resilience on the part of those most likely at risk of being "left behind." Thus in Zoom's "Future of Education" report, it can simultaneously recommend that "governments and institutions should prioritize equity of access for digital education" while at the same time asserting the need for a more widespread integration of educational technology within a broader, consumer-based "technology ecosystem" that will "enabl[e] students to learn in their own way, and with their real-world contexts" (26).

We see a similar convergence of "access" and productivity in the "Lessons from the Pandemic" report from the Information Technology and Innovation Foundation (ITIF)—whose mission is to "formulate, evaluate, and promote policy solutions that accelerate innovation and boost productivity to spur growth, opportunity, and progress" (ITIF n.d.). The report, written in the summer of 2020, simultaneously praises the US infrastructure's ability to support the massive increase in bandwidth demands as schools and businesses went online, while at the same time highlighting the gaps in rural access and affordability barriers for low-income users and the need for more to be done to bring about "a more just and effective broadband network for all Americans" (Brake 2020). What the Covid-19 pandemic revealed, as both Cabalquinto and Pyatovskaya suggest in their case studies in this volume, is the degree to which those most in need of services and support experience the greatest degree of digital precarity. As Kennedy, Holcombe-James, and Mannell (2021) note, access to data and devices provides a basic threshold for participation, but the ability to deploy these tools and orient oneself toward these sorts of engagements suggests a level of fluency beyond what many high-risk/high-need populations may already possess— often due to purposeful marginalization of these communities to begin with. But calls for greater access ultimately support the same ideological framework that equates technological access with social equity. Access reveals a disposition toward global networks and as such signals one's degree of social capital within a network society—a "state nobility" (Bourdieu 1996) for the digital age.

But of course the Zoom classroom was anything but a "frictionless learning environment"; students struggled to meet the performative criteria for what "engagement" was supposed to look like and, as we discuss in our own chapter, found strategies of resistance that introduced their own mode of data friction, from mild expressions of institutional critique through "Zoom University" memes (Chaudhry 2021) to outright refusal to "engage" in online classes by muting mics and turning off cameras. While the technological and ideological apparatuses that supported a logic of Zoom maintained a dominant structure that allowed the *flow of knowledge* to continue in the midst of crisis, it likewise offered if not modes of outright resistance to that logic, then "transversal" (Guattari 2015) lines of power that cut across those dominant structures. All three chapters in the section of this collection entitled "Transverse Networks and the Neoliberal University" explore how friction within a logic of the "frictionless university" gives rise to unsettled moments between control systems and the "outcomes" they attempt to order, be that through shadow networks, counter-assemblages of precarious parties, or unforeseen and unpredictable ecologies for teaching and learning.

Performing with and for Zoom

Very quickly in those early days of the pandemic, it became apparent—to quote from the title of a *New York Times* piece from March 17, 2020—that "we live in

Zoom now" (Lorenz et al. 2020). Daily life gained an increasingly hybrid quality, with Zoom framing what it meant to interact, and, in doing so, determining the conditions in which we performed a host of social and cultural functions. The implications of videoconferencing platforms becoming the de facto environment for social interaction reach far beyond the realm of interface design. In *The Presentation of Self in Everyday Life*, Erving Goffman (2008) explores how social actors move through their social environments, managing their identities in response to the space in which they find themselves and the audience (who are also social actors) within those spaces. For Goffman, the social environment provides the primary context for how and why social actors behave the way that they do. Goffman further denotes different spaces where our performances may shift: from public settings to smaller audiences, to private spaces where we can inhabit ourselves without any performance demands. The advent of social media, however, has added new layers to how we understand performance, audience, and public and private spaces. Indeed, Goffman's assertion that we are constantly managing our impressions feels particularly accurate when considering the added pressures of managing our identities in multiple social spaces—professional and personal—both face-to-face and online. With new demands in performance, competing imagery of other Zoom attendees with our mirrored reflection (Sumner 2024), and the uncanny appearance of human behaviors that may be muted, delayed, or even frozen, we are still navigating the complex micro social interactions and interpreting human behaviors in different social settings, even digital ones. Thus, when the world shut down during the Covid-19 pandemic, and all forms of social interactions shifted to digital spaces, the performative demands of working from home became all the more complex in the sharp merging of public and private spaces.

Privacy management was a near-constant narrative as we began asking: How much of our homes are we required to put on display to our classmates, coworkers, and even our friends? The hyperdependence on Zoom interactions forced an entry into the spaces that we so often kept private, leaving our social performances permanently on display. Prior to Covid-19, the networks of everyday life had already produced rather porous boundaries between public and private life, but for the most part, individuals managed to maintain some sort of partition between domestic, intimate spaces and their public performances of their professional and civic selves, if they so chose. It was a startling and humorous exception in The Before Times, for example, for a college professor to be interrupted in the midst of his *BBC News* (2017) interview by his children wandering into the room; the suspended possibility of the private irrupting in the midst of the public (or vice versa) haunts all of our network interactions, yet the exceptionality of these moments prior to life on Zoom speaks to how fully we ignored that fact and how successfully we managed to sustain an illusion of two distinct stages for social performance. Now, however, what was once the exception has become the rule. As millions of individuals found themselves Zooming from home while engaging coworkers, clients, patients, and students in professional interactions, the interpenetration of the public and private became a matter of daily fare. And yes, while early on in the pandemic several newsworthy (or at least meme-worthy) stories circulated widely, serving as videoconferencing cautionary tales—usually involving sex, drugs, or bowel movements—moments of transgressive privacy very much became the norm: we found ourselves, in the midst of the workday, peering into backgrounds of bedrooms and kitchens, examining decorations and personal effects, and sharing in the comings and goings of pets and other family members entering and leaving the frame. Some users opted for background images or made use of blurring effects to "hide the mess" of their daily lives. Others, however, seemed to embrace the transgression itself, implicitly or explicitly accepting the everydayness of this new liminality between public and private life.

And while we acknowledge the transgressive nature of these incursions of the domestic and the intimate into the frame of the Zoom workplace, these crossings also took place in the opposite direction, as spaces once designated as private became de facto workplace settings, and falling under the purview of a whole range of corporate and institutional policies that dictated appropriate and inappropriate behavior. As Howard notes in this volume (and as discussed earlier), these policies frequently had disproportionate impact on those individuals who were already under heightened pressure to maintain a clear (and visible) partition between workplace performance and personal life, based on race, gender, and class (see also Archer et al. 2024). We return to Professor Robert Kelly's performance on the BBC back in 2017, gestured to earlier in this chapter, to note that when the clip of his children intruding into his live, on-air BBC video call went viral, it did so framed as a "cute" disruption within an otherwise professional moment; what is commented upon less frequently is the highly gendered nature of this scene and in particular how—after his two children enter the frame—Kelly's wife is seen desperately trying to gather up her children and corral them out of the room, all the while crouching in a futile attempt to remain unseen. This brief irruption of the domestic into the professional frame of a 2017 live news broadcast serves as an apt allegory for the lived, daily experience of millions of women during the pandemic, forced through the penetrating gaze of Zoom to scramble to keep their domestic roles out of frame, all the while performing in their professional roles and contexts.

Quite literally, then, Zoom and other videoconferencing platforms served as technological and ideological apparatuses of surveillance. The panopticon's design ensured a sense of constant surveillance by obscuring the watcher from the watched, leaving subjects to assume—but never confirm—that they were under the scrutiny of their observers. Foucault (2020) describes the prisoners subjected to the panoptic gaze as "docile bodies," disciplined through surveillance but with no power to turn the gaze back toward the structures of power that infiltrated their existence with such invasive intent. Certainly, elements of a panoptic gaze assert themselves within the daily practices of videoconferencing and often do so in ways that blur the line between disciplinary surveillance and what are presented as "best practices" for attentive business team members and engaged online learners. As Finders and Muñoz (2021) note in an *Inside Higher Ed* article written one year into the pandemic: "Consciously or unconsciously, the need to have cameras on—while considered by many instructors as pedagogically sound—is actually indicative of an attitude toward teaching that positions students as docile bodies in need of constant surveillance." But with Zoom, however, it is increasingly unclear who is subjecting whom to the panoptic gaze, as those leading classes or meetings must also subject themselves to the watchful eye of those who observe their professional performances, including their own self-directed gaze. The implications of this literal self-surveillance (as opposed to the internalized self-surveillance that the panopticon instills) are not to be taken lightly as "we are forced to *see ourselves being seen*" (Sumner 2024, 194). Many online participants, especially in the classroom setting, have pushed back against these intrusions with cameras and audio turned off, thereby leaving the presenter—in an inversion of the panoptic apparatus of power—with a literal "black box" audience of blank screens and no indication of whether or not real observers were present behind their monitors. In these unstable digital spaces, we vacillate between observed and observer, visible and invisible, controlling subject and controlled object. As Bucher (2018) notes, under these new regimes, power exerts itself less as a "threat of visibility" than as a "threat of *invisibility*" to the extent that unstable digital subjects only coalesce as performative actors through their operative function within a network of flows (84–5). Thus, "participatory subjectivity is not constituted through the imposed threat of an all-seeing vision machine but, rather, by the constant possibility of disappearing and becoming obsolete" (Bucher 2018, 92).

Yet we should not lose sight of the fact that the *real* power of the webcam is less about what it sees than what it *extracts* for useful measurement. Software used for proctoring students who were forced to take exams from home became the subject of much debate in many campus communities, as students could no longer leave cameras off—or even step away from their computers—during exams lest the software mark them as cheaters (Flaherty 2020). Many workplaces also introduced similar tech to monitor eye movement, mouse clicks, and presence on screen (West 2021) to ensure that those working from home were really performing their duties as peculiar commodities by being *productive*. Thus, even when the option was present to turn off one's camera and mute oneself in an attempt to maintain some boundary between the public and private, software such as these programs became more commonplace to monitor activity and move the public into the private, ensuring the extractive gaze of data colonialism (Couldry and Mejias 2019) was not interrupted, even during a pandemic. In an already data-dependent era, more privacy and personal data has become available than ever before through online monitoring and the constant use of videoconferencing in work and social interactions. Such incursions of informatic biopower require further consideration within an emerging discussion of digital capital.

Closures and Openings

While Zoom became the default platform for a wide range of official and institutional practices, from corporate meetings to college class sessions, we witnessed during the height of social distance mandates and stay-at-home orders unanticipated engagements with the platform as well. The global pandemic hit many industries hard, but in particular, artists and performers—as well as their performance venues—saw a massive loss of space, audiences, and income. Many artists developed performance spaces through videoconferencing in order to maintain their practice and their connection to their audiences, while others developed new programs and worked to find accessible ways for community members to participate in online art programming. Thus, though performers were still faced with a grid of black squares as their performance venue, the invited gaze of a Zoom audience allowed for artistic performances to continue, whether as digital shorts, live streamed music sets, or isolated cast members performing many roles with a reduced cast list. Though the issue of access to the technology and bandwidth needed to partake in these live streamed events is still front of mind, the presentation of artistic performances through Zoom has allowed in many other ways for a larger audience reach, from those who may not live near a performance center to others who may not be able to access physical spaces comfortably or safely.

While commercial television and film needed to find ways to adapt production schedules to the reality of stay-at-home orders, with the Zoom interface often figuring prominently as a screen within the screen, the constraining logic of the platform likewise created moments of creative opening. While Zoom presents itself as a tool to keep a neoliberal economy flowing, it also provided a frame in which artists and audiences alike experienced novel spaces for performance. As burrough and Starnaman (2021) note in their discussion of a live, Zoom-based participatory event entitled "Epic Hand Washing for Synchronous Participation," which was performed during the 2020 Electronic Literature Organization (remote) conference, "As a result of the pandemic, people in technologically connected communities are intimately familiar with the online interactive public that was once the realm of digitally savvy producers and users. This reality thus broadens the audience for our online projects." What burrough and Starnaman explicitly refer to as the *playfulness* of their event speaks to a mode of engagement that attempts to find new headings, ones that may indeed disrupt expectations for efficiency and maximum performativity. Daniel Paul O'Brien's and Craig Fahner's chapters within this collection explore these potentials for creative opening within the context of artistic practice, but this same subversive potential expresses itself in institutional settings, such as in the use of Zoom's chat function as shadow network or "subcultural collective," topics raised by Heather J. Carmack et al. and Alexis-Carlota Cochrane and Theresa N. Kenney, respectively, in their contributions. Subversion of these network logics, however, need not map emancipatory practices, as has been the case in zoombombing— when uninvited guests find their way into unsecured meetings with the sole intention of creating disruption. On one hand, we can see zoombombing as a form of evil media practice (Fuller and Goffey 2012) that disrupts the dominant performativity logic of Zoom and undermines the assumptions of rational exchange that still drive much of how we understand "effective" communication. In practice, however, the dominant impulse embedded in zoombombing seemed more intent on closure of discourse and limiting potential explorations, most often in the form of racist, sexist, homophobic, transphobic, and/or anti-Semitic attacks (Nakamura et al. 2021; Ali 2021).

Structure of the Collection

We have divided this collection of essays into four parts and provided a brief introduction to the chapters included in each section. *Zoom Embodiments* explores the tensions in our experience of bodies and bodily affect in and through mediated video exchanges. These essays call attention as well to the interpenetration of public and private spaces and how mechanisms of surveillance can operate both to flatten individuals and to heighten social controls. *Staging Zoom* takes up the *place* of Zoom more explicitly and how the affordances and constraints of the medium impact both artistic and social performance. These chapters also suggest the potential for openings within the closures of the logic of Zoom, including novel ways of imagining what it means to engage and interact in an online environment. *Transverse Networks and the Neoliberal University* takes on "Zoom University" from the perspective of both faculty and students, with each of the three essays engaging in some form of autoethnographic work. They map out structures of power that Zoom both elides and enforces, while at the same time noting those moments in which subversion, subterfuge, and challenge express themselves as part of the same logic that seeks to ensure friction-free learning. And finally, **Unstable Connections** provides a discussion of precarious communication, both within marginalized communities and across a broad spectrum of users and usages. The concept of resilience returns in this section but in a more nuanced encounter, one that acknowledges the instability of human and technological networks, regardless of the technological promise of seamless, universal engagement.

Each of these chapters offers a thoughtful contribution to a burgeoning discussion on what Zooming means to us as academics, teachers, researchers, and community members. In an era of Covid-19, our relationships and experiences are deeply intertwined with our ability to Zoom. This shift resulted in new forms of artistic practice, new modes of pedagogy, and new ways of social organizing, but it has also created new forms (and exacerbated existing forms) of exploitation, inequity, social isolation, and precarity. Though investigations into the social effects of digital spaces are not new, this moment in time requires careful and critical investigation through the lens of a global pandemic as it intersects with a world that has never been more digital in its presence and social interactions. It is our hope that this volume provides a blueprint of sorts for other critical engagements and explorations of how our lives and digital landscapes have been impacted by Covid-19, regardless of the instability of our connections.

Notes

1   While the number of new, reported cases has continued to decline throughout 2023, the pandemic does indeed continue on. On November 13, 2023, the WHO reported 128,082 current, confirmed cases of Covid-19, with just over 772 million cumulative cases and nearly 7 million deaths (World Health Organization n.d.).

2   This being said, one of the authors of this piece, though well versed in all things tech, had barely heard of Zoom before the pandemic; this is not an unusual statement heard from many folks (and in educational settings in particular) and signifies again the impact that not only Zoom the company but also Zoom the term, the action, the tool, and the logic of life at home suddenly had on so many of us.

3   See also Deleuze's (1992) discussion of control as "modulation" and "continuous variation" (4–5).

4   While the wording has changed as of FY 2024, the "frictionless" sentiment remains the same in Zoom's (2023b) current corporate description as "an all-in-one intelligent collaboration platform that makes connecting easier, more immersive, and more dynamic for businesses and individuals. Zoom technology puts people at the center, enabling meaningful connections, facilitating modern collaboration, and driving human innovation."

References

Agarwal, Vivek. 2018. "The Frictionless Enterprise." *Forbes*, December 26, 2018. https://www.forbes.com/sites/forbestechcouncil/2018/12/26/the-frictionless -enterprise/.

Ali, Kawsar. 2021. "Zoom-ing in on White Supremacy: Zoom-Bombing Anti-Racism Efforts." *M/C Journal* 24, no. 3. https://doi.org/10.5204/mcj.2786.

Archer, Catherine, Marianne D. Sison, Brenda Gaddi, and Lauren O'Mahony. 2024. "Bodies of/at Work: How Women of Colour Experienced Their Workplaces and Have Been Expected to 'Perform' during the COVID-19 Pandemic." In *Performing*

Identity in the Era of COVID-19, edited by Lauren O'Mahony, Rahul K. Gairola, Melissa Merchant, and Simon Order, 140–61. New York: Routledge.

BBC News. 2017. "Children Interrupt BBC News Interview - BBC News." *YouTube*, March 10, 2017. https:// youtu.be/Mh4f9AYRCZY.

Beniger, James R. 1986. *The Control Revolution: Technological and Economic Origins of the Information Society*. Cambridge, MA: Harvard University Press.

Bourdieu, Pierre. 1996. *The State Nobility*. Stanford: Stanford University Press.

Brake, Doug. 2020. "Lessons from the Pandemic: Broadband Policy After COVID-19." *Information Technology and Innovation Foundation*, 13 July 2020. https://itif.org/ publications/2020/07/13/lessons-pandemic-broadband-policy-after-covid-19.

Bucher, Taina. 2018. *If... Then: Algorithmic Power and Politics*. New York: Oxford University Press.

burrough, xtine and Sabrina Starnaman. 2021. "Epic Hand Washing: Synchronous Participation and Lost Narratives." *M/C Journal* 24, no. 3. https://doi.org/10.5204/ mcj.2773.

Carey, James W. 1992. *Communication as Culture: Essays on Media and Society*. New York: Routledge.

Cardinal, Alison, Kirsten Higgins, and Anthony Warnke. 2023. "Valuing an Embodied Epistemology to Counter Neoliberal Programmatic Reform at the Two-Year College." *Journal of Basic Writing* 42, no. 1. https://doi.org/10.37514/JBW-J.2023.42 .1.04.

Chaudhry, Aliya. 2021. "Memes That Perfectly Sum Up the Horror of Going to Uni During a Pandemic." *Vice*, April 21, 2021. https://www.vice.com/en/article/v7m5ed/ zoom-memes-for-self-quaranteens-facebook-group.

Couldry, Nick and Ulises Ali Mejias. 2019. *The Costs of Connection: How Data Is Colonizing Human Life and Appropriating It for Capitalism*. Stanford: Stanford University Press.

Crano, Ricky. 2021. "Room Without Room: Affect and Abjection in the Circuit of Self-Regard." *M/C Journal* 24 no. 3. https://doi.org/10.5204/mcj.2790.

Dean, Jodi. 2009. *Democracy and Other Neoliberal Fantasies: Communicative Capitalism and Left Politics*. Durham: Duke University Press.

Deleuze, Gilles. 1992. "Postscript on the Societies of Control." *October* 59: 3–7.

Finders, Margaret and Joaquin Muñoz. 2021. "Cameras On: Surveillance in the Time of COVID-19." *Inside Higher Ed*, March 2, 2021. https://www.insidehighered.com /advice/2021/03/03/why-its-wrong-require-students-keep-their-cameras-online -classes-opinion.

Flaherty, Colleen. 2020. "Big Proctor." *Inside Higher Ed*, May 20, 2020. https:// www.insidehighered.com/news/2020/05/11/online-proctoring-surging-during -covid-19.

Foucault, Michel. 2020. *Discipline and Punish: The Birth of the Prison*. New York: Penguin.

Fuller, Matthew and Andrew Goffey. 2012. *Evil Media*. Cambridge, MA: MIT Press.

Gates, Bill. 1995. *The Road Ahead*. New York: Viking.

Goffman, Erving. 2008. *The Presentation of Self in Everyday Life*. New York: Anchor Books.

Goldberg, Emma. 2023a. "Even Zoom Is Making People Return to the Office." *New York Times*, August 7, 2023. https://www.nytimes.com/2023/08/07/business/zoom-return -to-office.html.

Goldberg, Emma. 2023b. "Return to Office Enters the Desperation Phase." *New York Times*, June 20, 2023. https://www.nytimes.com/2023/06/20/business/return-to -office-remote-work.html.

Grandinetti, Justin. 2022. "'From the Classroom to the Cloud': Zoom and the Platformization of Higher Education." *First Monday* 27, no. 2. https://dx.doi.org/10 .5210/fm.v27i2.11655.

Guattari, Félix. 2015. "Transversality." In *Psychoanalysis and Transversality: Texts and Interviews 1955–1971*, translated by Ames Hodges, 102–20. Cambridge, MA: Semiotext(e).

Information Technology and Innovation Foundation. n.d. "About ITIF: A Champion for Innovation." Accessed November 24, 2023. https://itif.org/about/.

Kennedy, Jenny, Indigo Holcombe-James, and Kate Mannell. 2021. "Access Denied: How Barriers to Participate on Zoom Impact on Research Opportunity." *M/C Journal* 24, no. 3. https://doi.org/10.5204/mcj.2785.

Kitto, Simon and Vaughan Higgins. 2010. *Pedagogical Machines: ICTs and Neoliberal Governance of the University*. New York: Nova Science Publishers, Inc.

Lemieux, Christiane and McDonald Duff. 2020. *Frictionless: Why the Future of Everything Will Be Fast, Fluid, and Made Just for You*. New York: Harper Business.

London Image Institute. 2021. "Zoom Meeting Etiquette: Dressing for Success While Working at Home." May 21, 2021. https://londonimageinstitute.com/zoom-meeting -etiquette-dressing-for-success-while-working-at-home/.

Lorenz, Taylor, Erin Griffith, and Mike Isaac. 2020. "We Live in Zoom Now." *New York Times*, March 17, 2020, Updated November 17, 2021. https://www.nytimes.com /2020/03/17/style/zoom-parties-coronavirus-memes.html.

Lovink, Geert. 2022. *Stuck on the Platform: Reclaiming the Internet*. Amsterdam: Valiz.

Lyotard, Jean-François. 1984. *The Postmodern Condition: A Report on Knowledge*. Minneapolis: University of Minnesota Press.

Marx, Karl. 1992. *Capital: A Critique of Political Economy*, Volume 1. New York: Penguin.

Morozov, Evgeny. 2013. *To Save Everything, Click Here: The Folly of Technological Solutionism*. New York: PublicAffairs.

Nakamura, Lisa, Hanah Stiverson, and Kyle Lindsey. 2021. *Racist Zoombombing*. New York: Routledge.

Nunes, Mark. 2006. *Cyberspaces of Everyday Life*. Minneapolis: University of Minnesota Press.

Pew Research Center. 2020. "How the Coronavirus Outbreak Has – And Hasn't – Changed the Way Americans Work." December 9, 2020. https://www.pewresearch .org/social-trends/2020/12/09/how-the-coronavirus-outbreak-has-and-hasnt -changed-the-way-americans-work/.

"The Power of the Pivot: Is "Pivoting" the Business Buzzword of the Year?" 2020. *CEO Magazine*, April 23, 2020. https://www.theceomagazine.com/business/management -leadership/pivoting-the-business-buzzword-of-the-year/.

Roberts, Laura Morgan and Courtney L. McCluney. 2020. "Working from Home While Black." *Harvard Business Review*, June 17, 2020. https://hbr.org/2020/06/working -from-home-while-black.

Roberts, Peter. 1998. "Rereading Lyotard: Knowledge, Commodification and Higher Education." *Electronic Journal of Sociology* 3, no. 3. https://sociology.org/ejs-archives/ vol003.003/roberts.html.

Roy, Alain J. 2022. "The Past, Present and Future of Remote Work." *Forbes*, December 29, 2022. https://www.forbes.com/sites/forbesbusinesscouncil/2022/12/29/the-past -present-and-future-of-remote-work/?sh=15052248cb2b.

Sumner, Tyne Daile. 2024. "Zoom Face: Self-surveillance, Performance and Display." In *Performing Identity in the Era of COVID-19*, edited by Lauren O'Mahony, Rahul K. Gairola, Melissa Merchant, and Simon Order, 184–98. New York: Routledge.

West, Darrell M. 2021. "How Employers Use Technology to Surveil Employees." *Brookings Institute*, January 5, 2021. https://www.brookings.edu/articles/how -employers-use-technology-to-surveil-employees/.

Wood, Elizabeth. 2023. "Tapping into the Future of Finance: 3 Experts Discuss the Ways Frictionless Payments Will Revolutionize Digital Banking Within the Decade." *Business Insider*, March 29, 2023. https://www.businessinsider.com/experts-explain -frictionless-payments-role-in-digital-banking-2023-3.

World Health Organization. n.d. "WHO Coronavirus (COVID-19) Dashboard." *World Health Organization*. Accessed November 24, 2023. https://covid19.who.int/.

World Health Organization. 2020. "WHO Director-General's Opening Remarks at the Media Briefing on COVID-19–11 March 2020." *World Health Organization*, March 11, 2020. www.who.int/director-general/speeches/detail/who-director-general-s -opening-remarks-at-the-media-briefing-on-covid-19---11-march-2020.

World Health Organization. 2023. "Statement on the Fifteenth Meeting of the IHR (2005) Emergency Committee on the COVID-19 Pandemic." *World Health Organization*, May 5, 2023. https://www.who.int/news/item/05-05-2023-statement -on-the-fifteenth-meeting-of-the-international-health-regulations-(2005)- emergency-committee-regarding-the-coronavirus-disease-(covid-19)-pandemic.

Zoom Video Communications, Inc. 2020a. "Zoom Video Communications Reports Fourth Quarter and Fiscal Year 2020 Financial Results." March 4, 2020. https://

investors.zoom.us/news-releases/news-release-details/zoom-video-communications
-reports-fourth-quarter-and-fiscal-year.

Zoom Video Communications, Inc. 2020b. "Zoom Reports First Quarter Results for Fiscal Year 2021." June 2, 2020. https://investors.zoom.us/news-releases/news-release-details/zoom-reports-first-quarter-results-fiscal-year-2021.

Zoom Video Communications, Inc. 2022. "The Future of Education: Lessons from Educators." https://explore.zoom.us/media/epub-lessons-in-education-v5.pdf.

Zoom Video Communications, Inc. 2023a. "Zoom Video Communications Reports Fourth Quarter and Fiscal Year 2023 Financial Results." February 27, 2023. https://investors.zoom.us/news-releases/news-release-details/zoom-video-communications-reports-fourth-quarter-and-fiscal-2.

Zoom Video Communications, Inc. 2023b. "Zoom Video Communications Reports Financial Results for the Second Quarter of Fiscal Year 2024." August 21, 2023. https://investors.zoom.us/news-releases/news-release-details/zoom-video-communications-reports-financial-results-second-0.

Zoom Embodiments

In this first section, we explore how the logic of Zoom foregrounds a paradoxical moment wherein the platform both denies and affirms the primacy of embodied experience.

Zoom offered a technological solution to pandemic-imposed isolation and, in doing so, revealed the degree to which neoliberal networks—technological as well as institutional and social—had already embedded themselves in the fabric of everyday life. While videoconferencing provides entry into a networked "space of flows" (Castells 1996) in which connectivity and circulation are synonymous with performance, at either end of the interface we still find human bodies situated before devices and screens. For all of their claims of creating "frictionless" work environments, Zoom and other videoconferencing platforms must still establish a material interface between mediated flows of data and embodied experience: the body as both terminal and node. To be "on Zoom," then, marks a scene of intra-action (Barad 2007) between bodies and networks, materializing the digital within an embodied space of the here and now, while at the same time framing material presence as a disembodied experience within an interactive digital display. While the causes of and cures for "Zoom fatigue" have been widely discussed in the months (and years) following those initial Covid-19 stay-at-home orders, one thing is clear: the malady itself serves as a diagnosis of this paradoxical-yet-banal experience of simultaneous embodied and disembodied presence. As Della Ratta (2021) notes, "it is precisely when acquiring a disembodied status that we are reminded of our flesh and bones, skin and muscles—by this enduring and endless mirror gaze."

It is understandable, then, why—in the midst of a global health crisis—missteps at the intersection of embodied and mediated presence (cameras and/or mics capturing "private" bodily functions) would become viral sensations, as screen recordings of hilarious and/or humiliating Zoom moments were shared across social media and mass media platforms alike, serving as videoconferencing cautionary tales through public shaming of bodily functions. After the novelty of Zooming from home gave way to what for many became a daily repertoire of back-to-back videoconferencing sessions, it became clear that this blurring of mediated and material presence would present an increasingly troubled and precarious terrain as transgressive, voyeuristic intrusions became the norm. Some users addressed this violation by embracing yet another technological solution for a technologically imposed problem, opting to use Zoom's "Background and Effects" setting to create a "virtual background" or a "blurring effect" to draw the digital blinds on their private settings. Others, however, seemed to accept the intrusion as more feature than glitch: an integral part of the platform vernacular for life on Zoom. (Gibbs et al. 2015).

And while domestic and intimate spaces offered scenes for intended or unintended transgressions within institutional norms, this border crossing occurred in the opposite direction as well, as spaces once designated as private became de facto workplace settings, and fell under the dictates of workplace and classroom policies for appropriate and inappropriate behavior, including dress, personal appearance, trespass of family members and/or pets, and the overall professional setting of whatever fell within the frame of Zoom. Perennial issues of race, gender, and class embedded in institutional coding for "professional behavior" were amplified through the lens and frame of videoconferencing platforms such as Zoom, as workplace and home space merged, placing the pressure to "manage" childcare, eldercare, and limited bandwidth (in every sense) on those individuals already within precarious relations to normative structures of power. And today, the consequences of these literal and ideological apparatuses of surveillance are still being played out in workplaces and educational institutions.

The chapters in this first section explore these challenges through three different approaches. Ricky D. Crano's chapter, "The Face of the Network: Subjectivity, Securitization, and the Production of Sad Affect on Zoom," provides an in-depth analysis of how the Zoom interface operates as a tool to maintain (and obscure) the extractive processes that are necessary to the smooth functioning of digital capitalism (Sadowski 2020; Zuboff 2019) by both insisting on a rhetoric of disembodied and dematerialized labor and at the same time placing great cognitive and corporeal demands on those who found themselves fettered to its network during the pandemic. Zoom fatigue, he argues, is deeply tied to these cognitive and corporeal demands, but it is also a symptom of the overarching weight of what he refers to as Zoom's "sad affect"—a draining away of potential and possibility. Core to this sad affect is a transformation of bodies into faces and more precisely the Zoom face (Sumner 2024)—a face of the network itself, in which we are all embedded and to which we all participate in supporting when we engage the interface and its logics.

In "Mediating the Death of a Parent: Zoom and the Deathbed Vigil," Susan A. Sci provides a powerful and deeply personal account of her Zoom deathbed vigil, along with other family members, as they attended their mother's final days. During the lockdowns in that first and most deadly wave of the pandemic in the New York City metropolitan area, Sci's mother contracted Covid-19 in her nursing home at a period in which any in-person contact between family members and facility residents was impossible. Sci's narrative discusses the nature of "the good death" when those deathbed rituals are no longer accompanied by the living, touching presence of loved ones and are instead mediated through a videoconferencing platform. Her account also asks us to consider to what extent individuals can appropriate a business efficiency tool such as Zoom for other, unanticipated purposes and to what extent that efficiency logic compels other practices to adopt that same imperative to *perform*. Sci's account poignantly reveals the lived negotiation of life under Zoom that is reminiscent of Nancy's (2000) "being singular plural" in her attempt to navigate a path between feeling a lack of "real presence" while still refusing to reduce this deathbed vigil to mere spectacle.

Jacquelyne Thoni Howard's chapter, "Zoom Etiquette Guides: Negotiating between Workplace Professionalism and Gendered Homeplace Surveillance in the Videoconferencing Borderlands," closes out this first section with a thoughtful critique of the demands of self-presentation on Zoom when social expectations cross into the digital/personal borderlands. Howard notes how the rapid move from work to home via videoconferencing extended the surveillance of employers into their employees' homes. Guidelines regarding "professional" behavior and appearance were suddenly applied to whatever fell within the frame of Zoom, regardless of the personal challenges and barriers many faced during the height of the Covid-19 lockdowns. These expectations were rooted in traditional workplace norms based on masculinity and whiteness, particularly regarding beauty ideals for women and specifically women of color. These work expectations were also rooted in classism and presented barriers for those caring for children or living in intergenerational family homes. As etiquette guides presented new rules for working over Zoom, Howard demonstrates how performative expectations in the Zoom borderlands only reinforced social inequities and systems of oppression present long before the pandemic began.

These three chapters set the stage for the larger issues at play in this volume and provide the groundwork for understanding this double articulation of embodied and disembodied experience, how the technological and ideological apparatuses of videoconferencing create and sustain these structures of surveillance, and the strategies of individuals who learned to work with—and against—these structures in their personal and professional encounters with the logic of Zoom.

References

Barad, Karen. 2007. *Meeting the Universe Halfway: Quantum Physics and the Entanglement of Matter and Meaning.* Durham: Duke University Press.

Castells, Manuel. 1996. *The Rise of the Network Society.* New York: Blackwell.

Della Ratta, Donatella. 2021. "Teaching into the Void: Reflections on 'Blended' Learning and Other Digital Amenities." *INC Longform,* January 6, 2021. https://networkcultures.org/longform/2021/01/06/teaching-into-the-void/.

Gibbs, Martin, James Meese, Michael Arnold, Bjorn Nansen, and Marcus Carter. 2015. "#Funeral and Instagram: Death, Social Media, and Platform Vernacular." *Information, Communication & Society* 18, no. 3: 255–68. https://doi.org/10.1080/1369118X.2014.987152.

Nancy, Jean-Luc. 2000. *Being Singular Plural.* Stanford: Stanford University Press.

Sadowski, Jathan. 2020. *Too Smart: How Digital Capitalism is Extracting Data, Controlling Our Lives and Taking Over the World.* Cambridge, MA: MIT Press.

Sumner, Tyne Daile. 2024. "Zoom Face: Self-surveillance, Performance and Display." In *Performing Identity in the Era of COVID-19,* edited by Lauren O'Mahony, Rahul K. Gairola, Melissa Merchant, and Simon Order, 184–98. New York: Routledge.

Zuboff, Shoshana. 2019. *The Age of Surveillance Capitalism: The Fight for a Human Future at the New Frontier of Power.* New York: PublicAffairs.

1

The Face of the Network

Subjectivity, Securitization, and the Production of Sad Affect on Zoom

Ricky D. Crano, University of California, Irvine

Introduction: Zooming, Feeling, Screening, and Other Pandemic Movements

"[I]t is absurd to believe that language as such can convey a message. A language is always embedded in the faces that announce its statements and ballast them in relation to the signifiers in progress and subjects concerned" (Deleuze and Guattari 1987, 179).[1]

Across much of the Global North, the first decades of the twenty-first century witnessed widespread popular excitement for digital communications dissolve into myriad laments for all that was missing from text-based talk (bodies, faces, looks, gestures, tones of voice) and grave warnings of social fragmentation, political polarization, and emotional stultification at the hands of Big Tech.[2] Early exhilaration at internet anonymity had by and large given way to social media incentives to cultivate and express one's true self (Facebook's "real name" policy) and moral panic around those who refused. Videoconferencing platforms like Microsoft's Skype (for desktops 2003, for phones 2013) and Apple's FaceTime (2010) promised to rehumanize conversation by giving it a face,[3] but the technology was impractical in most everyday scenarios and unnecessary wherever physical congregation could take place. It would take a pandemic-charged global lockdown for the demand to fully congeal. Zoom, founded in 2011, had 10 million daily users averaging around 100 billion annualized meeting minutes at the end of 2019; by April 2020, with Covid-19 infections rising rapidly worldwide, 300 million daily users were clocking 2.6 trillion annualized meeting minutes, a number that dipped slightly that summer before climbing to 3.3 trillion minutes by the end of the year (Zoom 2021a, 2021b). For students, teachers, and all those whose class position prioritized creative, cognitive, communicative, and managerial tasks over physical ones, *to zoom* had become a common verb.

In this chapter I treat Zoom, Inc. at once as a generic term for the sort of technocultural mediation it enacts, as a synecdoche for the videoconferencing economic space, and as the actual, profit-hungry publicly traded corporation that it is. As a visual medium, Zoom habituates users to exclusively flat, front-on views of each other's faces; exposes its participants to incessant, multiplicative eye gaze; and incites anyone with their video on into a spiral of self-reflection. Sonically, the platform deploys algorithms to suppress environmental noise and distends conversation through the introduction of compression and transmission lag. As a capitalist enterprise, Zoom models techniques of abstraction, datafication, and rent extraction pioneered decades earlier by the financial services sector and which are now hallmarks of venture-capital-fueled hi-tech.[4] Putting this together, I see Zoom, in its standard business and educational usages, and especially through the rolling lockdowns of the pandemic, as a means of amplifying class division and augmenting governmental programs of social control. In those classic disciplinary sites of the workplace and the school (Foucault 1977, 135–69), Zoom helps "unbundle" and assetize economic and academic labor by rendering communication more abstract, not so much through processes of disembodiment but through an intensification of certain corporeal pressures and an inducement to heightened states of self-control.

In its interrogation of the intersection of affect and work on the platform, this chapter dialogues with recent theoretical and experimental literature, across many disciplines, on the widely reported phenomenon of "Zoom fatigue." Through a close analysis of the face on—and *of*—Zoom, I advance a case study in the relationship between crisis capitalism and what I refer to as sad affect, or, more specifically, in the way firms can capitalize on a moment of social and economic crisis by extracting rent on the infrastructure augmenting people's massively diminished capacities for movement, touch, and affection. One's habituation to the Zoom meeting perpetuates familiar trends of the digitization of communications and culture, which Berardi (2011) analyzes through a rubric of "desensibilization," an inurement to other bodies that he finds rooted in the disjunct between infinitely expansive cyberspace and the inevitably delimited, increasingly scarce lived time of human attention (285).[5] In the Zoom room, one sees other spaces without really hearing or feeling them; the faces and voices therein are "lifted out" (Lash 2002, 21), as actual room resonance and background noise are algorithmically sifted away. Alone in front of the laptop camera, one is trapped in silence (should they be privileged enough to have a quiet, private space) and locked in place directly facing the screen; walk away, or fail to direct speech sufficiently toward the microphone, and the message disintegrates, tethered as it is to the presence of a face—even one with its video turned off. Gestures distill into generic icons and emoji; broad bodily movements are reduced to micro expressions, clicks, and swipes; three-dimensional spaces flatten onto screens; human heads flatten into faces; *distance*, that watchword of public health dominating pandemic discourse and behavior, does not collapse so much as it irrupts into the bodies at each end of the transmission, metastasizes, swells up inside them. In other words, one might think of Zoom along the lines of Guy Debord's (1995) analysis of "spectacle," defined "at once as society itself, as a part of society, and as a means of unification . . . [where] the unity it imposes is merely the official language of generalized separation" (12).

Outside of the corporate context, common uses of the word *zoom* imply a distance crossed—whether referring to a body or vehicle in motion or to a camera pulling into or broadening out from its subject—and often connote a sense of exhilaration at such crossing. In a great irony of its branding, Zoom the platform forecloses its eponymous action, as it immobilizes bodies and warns off movement. This, as I will argue, is the first of several touchpoints for sad affect to take hold in a Zoom meeting: the platform's promise to make distances irrelevant relies on techniques that shunt out closeness as well.[6] As elaborated in this chapter's final section, sad affect—even, and perhaps especially, in the form of partial joys unveiled as such—contributes to a diminishment of the capacity of subjects and, in this case, an inverse amplification of the corporate dominance of everyday life in the digital age. During pandemic closures, the mediation of separation by Zoom and its ilk was deemed essential, as offices and schools scrambled to patch together a semblance of community. Beyond that, Zoom affords an instrumentalization of communications in the classroom and on the job, a controlled socialization in sites of learning and office work, and a further disaggregation and assetization of the various types of labor undertaken therein. Diverse spaces become homogenized so that all gatherings take on the look and feel and functionality of the corporate meeting. Like all electronic media, Zoom promulgates what Lash (2002) calls an "atopic info space-time" that accelerates communication while promoting corporeal stillness, as if moving and speaking were mutually exclusive activities (147). It thus inverts the received sense of the verb *to zoom*, as it recodes the body, abstracts away positive affective charge, and promotes an insipid aesthetic of atomized talking heads, rigid grids, noise suppression, and facile customization. Along the way, the screen experience becomes increasingly dominated by a preponderance of talk. During the desolate stretches of the Covid-19 pandemic, as students and office workers acclimated to the look and feel of the platform and to the attendant routines of working and learning and socializing from home, they also came to experience the screen less as an aesthetic surface and more as a channel of multimodal communication, less a vehicle for the circulation of affective audiovisuality and more one for the circulation of verbal, visual, and text-based information.

In the 1970s, feminist activists and academics began reworking Marxist analysis to include housework and other socially reproductive chores as essential ingredients in the generation of surplus value and thus as urgent sites for the waging of class struggle (Dalla Costa and James 1972; Weeks 1998, 75–99). My approach in this chapter marries a Marxist-feminist attention to communicative, emotional, and domestic labor—in this instance, the *work* of seeing and listening and conversing on Zoom—with a theory of affect rooted in a Spinozan ethic of powers and embodiment. I follow scholars like Justin Grandinetti (2022), who, taking aim at "the highly problematic surveillant dimensions of partnerships between higher ed and big tech," clarifies that "COVID-19 did not 'cause' higher ed to rely on videoconferencing platforms like Zoom. Instead, the global pandemic functioned as crisis that served to accelerate already existing trends of platform, infrastructural, and capitalist surveillance" (2022). I endeavor to show how Zoom serves the interests of the neoliberal socioeconomic status quo—optimized workers, contented consumers (of commodities and words)—by decoding and recoding bodies and environments in its specific production and deployment of the face, what Gilles Deleuze and Félix Guattari (1987) describe as the work of a machine of "faciality" (167–91).

While Zoom, as I recount in more detail later, disturbs deep-seated conversational habits and distorts capacities for self-reflection, its formal constraints and basic audiovisual setup are largely on trend with the broader contours of the algorithmically mediated scopic field in liberal capitalist societies: a surge in egoism buttressed by a recalibration of otherness in terms of degrees of the same—an occlusion of "intersubjective lifeworlds" (Crary 2022, 92) by rapidly accumulating instances of communication, networking, and data-based connection. As a social salve through peak pandemic times, Zoom gave each of its meeting participants a perpetual project to work on—*the communicative powers of their own face.* Much as cryptocurrencies and no-bar-to-entry stock trading apps ostensibly "democratize" finance by giving anyone with a smartphone an excuse to be constantly checking the value of their accounts, Zoom, with its fixed close-up format and default "self-view," encourages constant examination of one's own countenance and physiognomy. The platform follows, first, reality television and, second, the social media selfie in similarly "democratizing" the face, that is, in sustaining an ever greater expectation of being seen and a concomitant need to actively, incessantly concern oneself with the crafting and cultivation of one's own brand and, with it, one's own visage (Crano 2021). As a corporate platform that immobilizes bodies, normalizes surveillance, and makes more work of everyday conversational exchange, Zoom contributes to the authoritarian drift of datafied cultures, reinforcing the class composition and power dynamics of late capitalist societies by dampening human capacities while expanding the supremacy of digital networks and those who own their components, control their supply chains, tap their flows, and extract rent on their manifold everyday uses.

Invisible Labor, or Platform Capitalism and the Production of Fatigue

Zoom offers to patch over the social and economic predicaments pursuant to the enactment of public health mandates, but its promise hinges on its users sustaining a burdensome amount of unpaid and underacknowledged attentional labor. Almost as soon as white-collar denizens of the pandemic-stricken Global North had learned to comport themselves to the norms and protocols of videoconferencing, reports of "Zoom fatigue" began pouring in.[7] Scholars across the disciplines have set themselves to work diagnosing the problem. The ophthalmologist Burton Kushner (2021), for example, attributes this fatigue in part to one's exposure to the "persistent eccentric gaze" of multiplex pairs of eyeballs darting around the screen, looking at once focused and askance, scanning for some semblance of eye contact (175). Cultural scholar Tyne Sumner (2022) expands on this observation, explaining how such a "distorted visual matrix" forces users to "perform new modes of engagement" in order "to create the illusion of intimacy and mutual understanding" in venues where the traditional markers of such sympathy have been rendered inoperable (870). Sumner (2022) contends that faces on Zoom, including one's own, are rendered "both hypervisible yet estranged," resulting not only in greater difficulty making sense of others' nonverbal signals but also in a need for participants to "*over*-perform [their own] expressions and gestures" (870). Furthermore, as the experimental psychologist Julie Boland (2022) and colleagues show, unpredictably variable lags in transmission "disrupt the rhythm of conversation" by throwing the "automatic neural timing mechanisms" that govern turn-taking in face-to-face exchanges completely out of sync, resulting in elongated transition times and fewer, longer conversational turns (1279). This contributes to a gradual deformation of subjectivity or what communications scholar Jeremy Bailenson (2021) describes as a transfiguration of users into managers—of conversation and of themselves—with every participant in a meeting now responsible for performing, deciphering, and negotiating the complex of new nonverbal cues needed to lubricate the machinery of platformized speech (3). Just to start a conversation demands a whole suite of logistical and infrastructural actions; it is little wonder that "fatigue" so swiftly sets in.

As opposed to mere tiredness, "fatigue" denotes a cause in some sort of physical or mental task-based exertion.[8] In the context of videoconferencing platforms, fatigue might be examined through a Marxist-feminist lens, which, at its most basic, illuminates the shifting class compositions and power dynamics resulting from the rapid evolution, organization, and exploitation of care work and communicative labor.[9] Such a framework would consider all of the previously cited factors contributing to Zoom fatigue, whether in a corporate or an academic setting, in terms of uncompensated labor required to maintain the (now virtual) sites of social reproduction and keep the communicative gears of the white-collar economy turning. All this extra work the platform creates—of continuous self-surveillance; of the overperformance of gesture and expression; of setting up, managing, and policing of meetings; and of psychosocial adjustment to disjointed conversational rhythms and the scattered, multiplex gaze accompanying interlocution—points toward exhaustion, malaise, and generalized feelings of disempowerment exacerbated by the worsening wealth disparities wrought by pandemic response policies.[10]

I situate Zoom fatigue—and the contributing factors to its production—within the domain of theory of affect, taken in the Spinozan sense of a relative increase or diminishment in the capacity or power of a body (which Spinoza, contra Descartes, sees as one with mind). Zoom fatigue, in essence, amounts to negative or sad affect generated through use of the platform. Much of work and school over Zoom is particularly depotentiating, since, with the dampening of individuals' powers to affect and be affected by other bodies in the world, the powers of the network swell. During the pandemic, investors profited mightily as enterprise subscriptions to videoconferencing platforms peaked, with firms and institutions of all stripes having become dependent on yet another outsourced infrastructural service, beholden to, among other things, its further rationalization of the workplace, its seamless interoperability with other popular corporate productivity and data management tools, and all the usage metrics it supplies.

Zoom is thus a portal onto the subjectifying forces of data capitalism and the commercial internet. It promises to enhance real-time telecommunication by interposing a face, yet the face it fashions is not a human one but the face of the network, of the internet itself and all the bodies it comprises. In the spring of 2020, with pandemic controls in full force and so many starving for face-to-face exchange, videoconferencing platforms gave their participants something else entirely by placing them *face to interface*. In a standard Zoom meeting, one chooses to view either one "spotlighted" face at a time (typically that of the speaker given the floor, or in many cases simply the loudest speaker, as determined by the spotlight algorithm) or a grid displaying everyone in the meeting all at once. Regardless of one's view, the faces one encounters on the screen, including one's own, appear, in Sumner's (2022) words, "estranged . . . uncanny and alienating . . . most unlike the faces we see in the supermarket or in passing on the street" (866, 869, 872). The eyes and mouths one follows in navigating a conversation repudiate the automatic and autonomous functions of organismic existence. Instead, on Zoom, one is confronted with a face or a series of faces that appear almost exclusively front-on and up close, cast against a customizable backdrop, abstracted from the body and even from the head, there or not there at the click of a button, often "muted," sometimes glitching, and always competing with a perpetual incitement to "self-view."[11] I find Deleuze and Guattari's (1987) concept of faciality instructive here, as it highlights the face—at the intersection of language and power—as something modular, machinic, disembodied, and veritably inhuman.[12] Zoom and its cohort of videoconferencing platforms take this regime of faciality to a new level; in so doing, and because of the way the platforms expand some capacities of the face while limiting others, they ultimately serve to scaffold the participatory authoritarianism endemic to commercial internet culture as it has evolved in neoliberal and liberalizing capitalist societies, which have coterminously—and, as I argue elsewhere, not coincidentally—spawned burgeoning fascist and white supremacist movements and elected scores of corporatist-populist demagogues to high-profile political posts (Crano 2022).[13]

Faciality Machine: Manufacturing the Zoom Subject

To clarify, it is "not a face" that Deleuze and Guattari (1987) are most interested in "but the abstract machine that produces faces according to the changeable combinations of its cogwheels" (168). This "faciality machine"—a "white wall-black hole system"—inputs bodies and lifeworlds, "decodes" and "overcodes" them, and outputs "signifiance" (defined as the material flux of meaningful communication) and "subjectification" (defined as the measure of a body's potential relative to others) (Deleuze and Guattari 1987, 170, 181). It is not the face itself but "the social production of the face," which anchors in language the relationship between subjectivity and power (Deleuze and Guattari 1987, 181). At the beginning, the faciality machine is what allows infants to grasp: (1) that there's a subjectivity behind their food source, revealed in the "black holes" of eyes and mouths, and (2) that they possess the power to signal to that subjectivity their desire for food; the "white wall" of the face provides a surface for inscription or a screen for the display of signs, a target for the infants' cries. It is a trap (Cochoy 2007). "The white wall of the signifier, the black hole of subjectivity, and the facial machine are impasses, the measure of our submissions and subjections"—first and foremost to the all-too-human necessity of shunting desires into language and to the particular assemblage of power that this semiosis implies (Deleuze and Guattari 1987, 189).

Facialization thus begins with a beheading of sorts. As Deleuze and Guattari (1987) put it, a face can only appear "when the head ceases to be a part of the body, when it ceases to be coded by the body . . . [when the faciality machine] removes the head from the stratum of the organism, human or animal, and connects it to other strata, such as signifiance and subjectification" (170, 172). The face "displaces [the body's] multidimensional, corporeal code," dampening its sensitivities and flattening its expressions; it in turn transmogrifies "all its surroundings and objects," making unruly "worlds and milieus" into serviceable "landscapes" or, in the era of Zoom, customizable backgrounds (Deleuze and Guattari 1987, 170, 181).The videoconferencing platforms that became almost instantaneously ubiquitous at the onset of Covid-19 lockdowns epitomize and extend the ongoing project of facialization, augmenting corporate stratification, workplace segregation, and classroom alienation as they optimize the efficacy of institutional communication through the telematic production, circulation, distortion, and arrangement of faces. The standard Zoom face—flatly front-on and fairly close-up—follows the formal template of the selfie and the social media profile pic and makes speaking and listening feel like any other online interaction. In this mode, more concentrated, intentional communicative performances are paradoxically coupled with a proliferation of messages that arrive increasingly detached from discernible human subjectivities: applause emoji that float across a Zoom webinar display, a voice that cuts in from an accidentally unmuted and invisible seminar attendee, a disjunctive collective gaze, and so on (Bailenson 2021, 2). As it simultaneously streamlines and contorts workplace and classroom communication, Zoom enhances the faciality machine, pruning back participants' powers to affect and be affected by others by presenting them with a screenful of nothing but faces looking anywhere and everywhere all at once.[14]

Despite the appeal of and to a more complete means of participation, Zoom tends to produce a subjectivity that is *subject to* and *through* logics of corporeal discipline and infrastructures of incorporeal affect. That Zoom (like Microsoft Teams and others) calls its alternative to the spotlight option the "gallery view" speaks to a spectatorial passivity presumed of the videoconference meeting-goer and reinforced by the platform.[15] Consider, for example, what is perhaps the most common "gallery view" of a medium-sized meeting, of a dozen or a few dozen people (say, of an office team, a college class, a volunteer organization, or an activist group, which all become structured according to the model of the corporate "meeting"): One would likely see a grid of participants, some with video on and some with video off. Boxes with video on would display faces, with discrete backgrounds for each. Those with video off would appear as black boxes, containing either a profile photo or an avatar or a color block with sans serif initials of the meeting participants. Sometimes the grid looks more like a checkerboard, as, owing to the predilections of the arrangement algorithm, those who flip their video off briefly will reappear in a different location of the grid. Someone on mute could be speaking in the box next to an unmuted speaker with their video off. The "talking head" in charge of a meeting might freeze up or cut out at any moment unannounced. Under the spell of extreme facialization, where significance and subjectification proceed *even in the absence of a visible human face*, meeting participants must adjust to emerging corporealities, expectations, and norms of communication that continue to decode the body and diminish the capacities of the individual while rerouting unpredictable aesthetic encounters into narrow, secure, proprietary bands of information exchange.[16]

Given its promise to sustain social harmony and economic efficiency through an extended public health crisis and waves of state-mandated physical isolation, the Zoom face synchronizes the coordination of libidinal and economic production, adjusting psyches and subjects to better serve what Sean Cubit (2016) calls "corporate cyborgs"—"huge agglomerations of technologies with human implants," "corporations . . . composed of nonhuman actors with human biochips embedded to carry out specialist tasks like those involving human resource management and public relations" (34). Such "human biochips" are the workers most likely to have retained their jobs through pandemic closures, either because what they do was deemed "essential" or because what they do can be remediated via platforms like Zoom. In promoting "safe" working conditions by affording disembodied and "disembedded" communication, Zoom crystallizes the iconicity of the face, in a crescendo of what selfies and reality television had achieved in prior decades: an incessant up-close exposure of and to the form of the face, the look and feel of which is shaped by the parameters of proprietary corporate infrastructure, not limited to compression algorithms, experiential design decisions, options for emotive expression, and access to the energetic and mineral resources required for saleable data storage and transmission.

The Black Hole of Computation

The Zoom face thus ensures the system compatibility of the organism. This is its contribution to the deepening securitization of society through algorithmic governmentality and surveillance capitalism (Amoore and Raley 2017; Bucher 2018; Cheney-Lippold 2017, 93–150; Zuboff 2019). The videoconferencing platform shores up subjectivities for new, digitally enhanced regimes of power and control. But neither the subject nor the face that fronts it should be confused for anything like an individual or the human. In fact, as Deleuze and Guattari (1987) put it, the face is, "from the start . . . [t]he inhuman in human beings . . . It is by nature a close-up, with its inanimate white surfaces, its shining black holes, its emptiness and boredom" (171). Consider how in the case of Zoom fatigue, the immobilization of the body, exposure to eccentric gaze, and disruption of the unconscious neural timing of conversation make the videoconferencing platform a ripe vehicle for the perpetuation of feelings of "emptiness and boredom" among meeting-goers, who come to see themselves—and their own faces—algorithmically operationalized in the name of optimized corporate communications.

Deleuze and Guattari (1987) continue: "It is not the individuality of the face that counts, but the efficacy of the ciphering it makes possible, and in what cases it makes it possible. This is an affair not of ideology but of economy and the organization of power" (175). Severed from their bodies, faces on Zoom lubricate the social machinery of capital. The "cipherings" they facilitate reveal a unified message of institutional power, or "spectacle," to Debord (1995), with its dream of "separation perfected" (11–24). Actual human faces themselves, on Zoom, become more like interfaces (Galloway 2013), where it is not so much what one sees that matters as the channels of communication and control that the surface opens up. As a matter of "economy and the organization of power," the face—here the Zoom face, the face of the network—is underwritten by a specific arrangement of bodies and knowledges on a global scale. To invoke an older usage of the "cipher," one could say that individual faces on Zoom are "zeroed out" in the process of overcoding a *corporate* or *network* body made up of minerals, chips, cables, data centers, e-waste dumps, and all the human hands and eyes and brains required to keep it all coordinated. The faciality machine recodes and covers over this composite body, securitizing the network by corralling attention around that which, deceptively, appears most human.

While the proverbial "face of the network" typically denotes a synecdochic figure whose appearance effectively conveys the character and values of a media brand, I deploy it here, in the context of pandemic-era Zoom and in consideration of the "network of networks" that is the internet, to demonstrate how a new wave of facialization advanced by videoconferencing platforms deflects care and concern away from matters like extraction, transportation, manufacture, and disposal by encouraging an understanding of digital environments as utterly dematerialized. The face of this network is not and cannot be that of an individual; it is rather the grid or "gallery" form itself—like the agency of the corporate cyborg, a "distributed but not communal" face (Cubit 2016, 34). Sitting at one's desk—face to *interface*—one struggles to look the right way, to locate any reciprocating gaze, to take any solace in the incompossibility of being stared at by all and none at once. What one sees are not other humans but a collection of faces that belongs exclusively and proprietarily to the platform. The screen supplies the "white wall" that delimits and arranges signifying flows; the checkerboard of black boxes and multiplex gazes combine with the cold opacity of the camera eye to constitute the "black hole" of a nebulous, inorganic subjectivity seeking to expand its own capacities and affective might.

For Deleuze and Guattari (1987), digital computation takes off from the same foundations as Western semiosis; for computer languages to work, "the black hole/white wall system must already have gridded all of space"; without it, "no message would be discernible and no choice [such as between a zero and a one] could be implemented" (179). The facialization at work in computation is evident in the simple fact that one opens their laptop or glances down to their phone with the expectation that—notwithstanding the incredibly complex history of extraction, manufacture, and transport behind it—the surface will bear signifying marks, messages, and potential meanings. The cyborg subjectivity behind the messages and choices of computation is continuous with the inhuman subjectivity the machine of facialization has always produced, and it is inseparable from the long-running projects of imperialism, colonization, militarism, racism, patriarchy, and unhinged economic development (Deleuze and Guattari 1987, 176).[17]

"Business process outsourcing" (BPO) names a prevailing trend of the digitally charged post-Fordist global economy, wherein problematic practices move increasingly to the peripheries of consumer markets and the margins of community concern. Zoom can be said to contribute to the BPO-ization of all data flows. Lisa Nakamura and coauthors, in their study of "racist zoombombing," show how, during pandemic lockdowns especially, videoconferencing software provided a platform for hate and aggression to white supremacist, homophobic, and misogynist bullies, pranksters, and trolls (Nakamura et al. 2021). I see Zoom even more broadly as a case study for the internet as a whole, a study in the primary tendencies of the commercialized internet and the datafication of social institutions: a corporatization of educational spaces and other sites of social reproduction; a normalization of surveillance; an assimilation of otherness into the orbit of the self; a homogenization of cultural forms; and a ravaging of the

Earth and its inhabitants through profligate energy expenditure and geological extraction.

Resisting the Sad Affects of Data Power

It would be understandable to see this as a thoroughly dispiriting affair. Affect theory, however, which hones the desubjectifying force of that which can be felt, transmitted, and shared at a nonrepresentational level, provides something of a road map for overcoming the "impasses" of the faciality machine amplified by Zoom. For the feminist cultural theorists who have done most to develop it (at least outside of the psychological, neurological, and cognitive sciences), the concept of affect affords a means to understand the intersection of the personal and the political without resorting to the standard liberal-individualistic framework of representation.[18] Affect involves and invests the body and links up closely with emotion but only insofar as emotion is understood to be socially, rather than privately, organized. Leading affect theorists like Eve Sedwick and Sara Ahmed take inspiration from Freudian psychoanalysis to highlight the centuries of neglect of sensation and embodiment by enlightened humanists and to demonstrate how the circulation of feelings like shame, anger, envy, and fear are made to serve the needs of the racist, misogynist, heteronormative, neoliberal status quo by cultivating dejected subjects with cramped capacities for experimentation and evolution (Sedwick 2003; Ahmed 2004; Ngai 2005). For Brian Massumi (2021), it is through affect that "the political becomes directly felt" (317) as an "inexpressible excess" (277) that specifies a body's "powers of existence" (315) or, in Spinoza's (2018) phrase, a "mind's power of thought" (141). Drawing on Spinoza more than Freud, Massumi (2002) describes the fundamentally transindividual and asignifying nature of affect—"not linked to an interior state or individuated subject," as Eugenie Brinkema (2014, 24) has it— as possessing an inherently liberatory or revolutionary potential. Affect inhabits what he calls a "real-material-but-incorporeal" (Massumi 2002, 5) dimension of experience, one that cuts against the grain of signifying flows and circuits of subjectification. Parting ways with the "linguistic model" that had held sway over post-structuralist cultural theory for decades, affect theory attends to language as it does to politics, that is, not in its representational capacity but as a material force, as something that gets deployed and wielded and which is inextricably entwined with relations of domination (Massumi 2002, 4).

Foregrounding affect can serve as a prophylactic of sorts for the Zoom subject. While Zoom partakes in the data-cultural trope of *dividuation*—that is, reimagining individuals in terms of fragmented quantities and attributes sliced up, rationalized, and recombined algorithmically across space and time (Deleuze 1992)—affect theory pushes in the other direction, toward an exploration of and opening onto *transindividual* powers in common. In Spinoza's ethics, while affection (*affectio* in Spinoza's Latin) refers to an emotion state, affect (*affectus*) is reserved for the passage or degree of difference between two states, registering as imperceptible "phase-shifts of the body" (Massumi 2002, 4) and an increase (via positive affect, i.e., Spinozan joy) or decrease (via negative affect, i.e., Spinozan sadness) in a body-mind's capacities relative to and in conjunction with others (Spinoza 2018, 95; Deleuze 1990, 93, 218).[19] Massumi (2021) reframes this ethics of joy and sadness, good and evil, in terms of the dueling concepts of power that have evolved in many Latinate languages, as in the French *puissance* (power as potential or strength, from the Latin *potentia*) and *pouvoir* (power as rule or authority, from the Latin *potestas*), which he translates respectively as "power-to" and "power-over" (315–16; cf. Deleuze 1990, 222–5).[20] Simply put, joyful encounters enhance one's inevitably embodied feelings of power and influence, while sad encounters diminish them. "Power-to is the power to change," Massumi (2021) says, as opposed to "power-over, which limits power-to . . . through . . . the normative channeling of activity" (316). Political authority is power over body-minds ("modes" to Spinoza), the abrogation of their freedoms and expropriation of their thought and action. As Massumi (2021) puts it: "Politics is the capture of powers of existence, turned against their own expansion and enhancement" (316).

In both its liberal and authoritarian forms, political power-over (*pouvoir*) is essentially a *media effect*, arrived at through both tactical media engagements (e.g., Mussolini's newspaper columns, Trump's social media) and trends in mainstream public discourse (ranging from *The New York Times* and *Fox News* to one's local social media feed). In the medico-political disaster that was Covid-19 through much of 2020 and 2021, hundreds of millions suddenly learned through videoconferencing to acclimate to new and intensified cultures of surveillance and self-control, forms of corporate policing and algorithmic authority, and structures of network homophily (Chun 2017, 2021). Zoom, as a media of social reproduction at an especially careful and controlling historical moment, functions as a power-over masquerading as a power-to, adopted en masse by enterprise clients eager to sustain workplace communications and camaraderie in order to protect their bottom lines. Like other popular media platforms that became more or less essential lifelines in the manifold (social, political, economic, public health) crises of the pandemic, Zoom promises the joys of expression but ends up, through its abstractions and algorithms and overcodings, separating individuals as much from themselves as from others. The result is a collective body not devoid of affect but one in which predominantly sad affect circulates and delimits its potential, its transindividual power-to, its exposure to otherness, its capacity to change.

One might call the general Zoom disposition a "sad joy" or "partial joy," along the lines of what Deleuze (1990) finds in Spinoza's early modern diagnosis of the human condition (245–6). The positivity of connection and expression is undercut by deepening social stratification, workplace atomization, and the compartmentalization of experience. Feeding into the well-established gigification of labor in companies and classrooms alike, Zoom (and many kindred corporate communications platforms and data services) obfuscates the relations between workers and bosses, maintaining strict physical separation while affording vast new capacities for managerial surveillance and control. A perpetual encounter with the face of the network diminishes human capacity by overcoding bodies into mere ciphers for social reproduction and economic efficiency. An ability to respond meaningfully to and with others, outside environments, risk, and chance are requisites for "change." Affective transmission needs a charge, a spark from other bodies—bearing in mind Brinkema's (2014) Spinozan reminder that "anything can be a body" (259). Those composite human-computer bodies one encounters on Zoom are *too like oneself* to spark any significant change. For example, those privileged enough to keep working via videoconference through pandemic lockdowns got the requisite face time with their teams without ever having to confront those other bodies—often racialized and gendered—doing custodial work, food service, infrastructure maintenance, and so on. When functioning smoothly, Zoom, at a macro level, sustains the communication and sociality essential to the maintenance and growth of capitalist institutions; at a micro level, it marks the apogee of digitally induced narcissism (Crano 2021; Han 2018, 40–3). In the Zoom room, the other is no object; all modes of nonself get squeezed out. Everyone has the same face.

The social model that comes out of this is a more face-forward, real-time, institutionally structured iteration of the "discriminating" homophilic networks Wendy Hui Kyong Chun (2021) theorizes as the disconcerting default form of online community in the age of the commercial internet. For Deleuze and

Guattari, the faciality machine plays a pivotal role in the cultural history of European racism and colonialism, which works, they insist, not through exclusions but by incorporation and arrangement according to degrees of sameness with the norm. The incorporative strategy of platforms and networks continues this trajectory, securitizing the self in its production of subjectivity and signifiance. The importance of Zoom gives an occasion to interrogate the long-standing but continually updated manufacturing of desire for more face. The abstractions of facialization occlude any view of the body of the network as a physical system of extraction and consumption and emission and waste. But affect theory teaches that subjectification is never final, that the hold of the network is ultimately quite fragile, its databases less secure and orderly than one might think.

Computers—and the subjectivities and signifying flows that route through them—have, as Deleuze and Guattari (1987) put it, an "exceptional need to be protected from any intrusion from the outside. In fact, there must not be any exterior" (179). The "web of subjectivities" held together by network computation is tenuous, always threatening to come undone should some "outside tempest" come to crack the "semiological screen" that makes possible the conveyance of digits through local machines, data servers, and clouds (Deleuze and Guattari 1987, 179). The apparatus of surveillance and control that Zoom helps prop up is far from totalizing, its datasets often proprietary and noncompatible, or else overloaded and unstructured. Zoom provides an apt look—a fitting face—for this messy assemblage. As with the social power of the financial sector, the data power of the network—its capture of subjectivities—hangs on its strength in measuring, predicting, and assetizing risk at increasingly granular levels, but it has little time for uncertainty. Writing in the 1920s, the economist Frank Knight (1921) first conceptualized the difference between *risk* and *uncertainty* in terms of the former's adherence to accountable and manageable probabilities and the latter's indexing of systemic change. Applying Knight's paradigm to algorithmic modeling, Justin Joque (2022) describes the inverse relationship that can inhere between risk and uncertainty, since "[u]ncertainty threatens every model, because there are always dangers that lie outside the closed space of the system . . . [T]he more efficiently systems manage risk, the harder it becomes to imagine and prepare for uncertainty" (39). Computers and networks, from this perspective, are fragile, demanding assemblages, incapable of processing any contingent event that would come "from the outside": a failure of the energy grid, a fire in the data center, or a flood through the cloud. Simply imagining such an event can put a spanner in the works of the Zoom-enhanced faciality machine, jar its subjectivities, and gum up its signifying flows. It is no coincidence that the imaginary is precisely what technosocial developments like algorithms and generative artificial intelligence, constituted and consumed as they are wholly by symbols, threaten to foreclose. This is but one site on which battle will be waged in the ongoing struggle for equality and justice in the digital era.

Notes

1 Brief portions of this chapter appear, in an earlier version, in my contribution to the "zoom" special issue of *M/C Journal*, edited by Mark Nunes and Cassandra Ozog (Crano 2021).

2 The psychologist-anthropologist Sherry Turkle's singular career trajectory exemplifies this evolving discourse, as her techno-optimistic cyborg ethnographies of the 1980s and 1990s made way for more public-facing, profoundly pessimistic studies of the ubiquitous digitality of the smartphone era (Turkle 1984, 1995, 2011, 2015).

3 In Skype's (2016) ad for its mobile product, for example, the company promises "an exciting and spontaneous way to stay connected," averring that "you can't beat talking face-to-face, seeing people's reactions and really experiencing that closeness with the ones you love." For its part, Apple's early FaceTime ads (ShonMStudio 2013), as well as CEO Steve Jobs's original product announcement at the Worldwide Developers Conference 2010, were mostly free of any narration or explanation and instead simply demonstrated FaceTime calls in action, suggesting that the product could speak for itself and that the desire for such a platform is innate (Steve Jobs Videos 2014).

4 On the role of rent extraction in financialization, see, for example, Harvey (2002), Hudson (2021), and Marazzi (2010).

5 "Sensibility," Berardi (2011) argues, "is in time, and space has grown to become so dense that the sensible organism . . . has no time to extract meaning and pleasure from the experience" (285). There is an unshakeable ethical component to this as well, as "[e]thics and sensibility have much in common. Ethical behavior . . . is based on the pleasure of the bodily presence of the other" (Berardi 2011, 285).

6 Here I take up a thesis of Byung Chul Han (2018): "Closeness and distance are interwoven, kept together by a dialectical tension . . . The abolition of distance does not create more closeness, but rather destroys it. Instead of closeness, a complete gaplessness ensues . . . Digital gaplessness removes all varieties of proximity and distance. Everything is equally near and equally far" (6).

7 The term "Zoom fatigue" started appearing in *The New York Times* and across the web by the end of March 2020. By the end of April 2020, reports on Zoom fatigue and lists of tips for avoiding it had appeared in sources ranging from *USA Today*, *Forbes*, and the *Wall Street Journal* to *Psychology Today* and *National Geographic*, among many others.

8 *Oxford English Dictionary.* "Fatigue," definition I.a.

9 See, for example, Cox and Federici (1976), Dalla Costa and James (1972), and Davis (1981). This work originates with a demand that Marxism attend to long overlooked categories of socially necessary labor such as unwaged housework. A notable forerunner to my own application of these methodological concerns, Lindsay Weinberg (2019) adapts a Marxist-feminist framework to critique cultural currents of algorithmic personalization and data-based domestic technology.

10 On the widening of the wealth gap during the pandemic, see, for example, Ferriera (2021).

11 Bailenson (2021) describes the self-view feature as a "constant mirror" that invites "self-evaluation and negative affect" (4). Sumner (2022) finds the self-view feature especially problematic as well, making the case that this "unprecedentedly regular encounter with our own face" makes it so that "interaction with other people becomes secondary to performing the self for our own individual validation" (866, 874). I have written more extensively about the psychodynamics of self-view on Zoom elsewhere; see Crano (2021).

12 As detailed in this chapter later, I draw primarily from the "Year Zero: Faciality" chapter in *A Thousand Plateaus* (Deleuze and Guattari 1987, 167–91).

13 This chapter follows Nakamura et al. (2021) in situating Zoom within the broader evolution of the internet as a tool of racist and misogynistic hatred. Tracking incidents of racist zoombombing during the early waves of the pandemic, they recount how "[t]he internet did not become a trashfire all of the sudden: it happened over a long period of time" (Nakamura 2021, 24). See also Nakamura and Chow-White (2012).

14 I take this to be more or less in keeping with Deleuze's (1986) thesis, at the end of *Cinema 2*, that the aesthetic image as it evolved through the first century of film was, at the time of his writing in the 1980s, gradually making way for an "information-image" in which screens swell with data and multimodal communications at the expense of formal castings of movement, perception, affect, and time.

15 The term "gallery," for example, calls to mind an art museum or the area reserved for onlookers in a chamber of public debate, both sites of spectatorship, that is, spaces wherein subjectivities are conditioned according to the formal and affective parameters of their looking.

16 I discuss this at length elsewhere; see Crano (2021).

17 On the socially oppressive military-industrial origins and development of electronic computing, see, for example, Edwards (1997), Franklin (2021), Katz, (2020), and Noble (2018).

18 See, for example, Sedwick (2003), Ahmed (2004), and Ngai (2005).

19 As Spinoza (2018) notes: "By *affectus* I mean affections of the body by which the body's power of action is augmented or diminished, assisted or restrained, and at the same time the ideas of these affections" (95).

20 Through much of this chapter, in an effort to bring more clarity to this distinction, I have followed alternative translators of Spinoza's Latin and Deleuze's French to render *puissance* as "capacity."

References

Ahmed, Sara. 2004. *The Cultural Politics of Emotion*. Edinburgh: Edinburgh University Press.

Amoore, Louise and Rita Raley. 2017. "Securing with Algorithms: Knowledge, Decision, Sovereignty." *Security Dialogue* 48, no. 1: 3–10.

Bailenson, Jeremy N. 2021. "Nonverbal Overload: A Theoretical Argument for the Causes of Zoom Fatigue." *Technology, Mind, and Behavior* 2, no. 1: 1–6.

Berardi, Franco. 2011. *After the Future*. Edited by Gary Genosko and Nicholas Thoburn. Oakland: AK Press.

Boland, Julie E, Pedro Fonseca, Ilana Mermelstein, and Myles Williamson. 2022. "Zoom Disrupts the Rhythm of Conversation." *Journal of Experimental Psychology* 151, no. 6: 1272–82.

Brinkema, Eugenie. 2014. *The Forms of the Affects*. Durham: Duke University Press.

Bucher, Taina. 2018. *If...Then: Algorithmic Power and Politics*. New York: Oxford University Press.

Crary, Jonathan. 2022. *Scorched Earth: Beyond the Digital Age to a Post-Capitalist World*. London: Verso.

Cheney-Lippold, John. 2017. *We Are Data: Algorithms and the Making of Our Digital Selves*. New York: New York University Press.

Chun, Wendy Hui Kyong. 2017. *Updating to Remain the Same: Habitual New Media*. Cambridge: MIT Press.

Chun, Wendy Hui Kyong. 2021. *Discriminating Data: Correlation, Neighborhoods, and the New Politics of Recognition*. Cambridge: MIT Press.

Cochoy, Franck. 2007. "A Brief Theory of the 'Captation' of Publics: Understanding the Market with Little Red Riding Hood." *Theory, Culture & Society* 24: 203–23.

Cox, Nicole and Federici, Silvia. 1976. *Counter-Planning from the Kitchen: Wages for Housework: A Perspective on Capital and the Left*. New York: New York Wages for Housework.

Cubit, Sean. 2016. *Finite Media*. Durham: Duke University Press.

Crano, Ricky. 2021. "Room without Room: Affect and Abjection in the Circuit of Self-Regard." *M/C Journal* 24, no. 3. https://doi.org/10.5204/mcj.2790.

Crano, Ricky. 2022. "The Joys of Following: Network Fascism and Micropolitics of the Social Media Image." *Deleuze and Guattari Studies* 16, no. 2: 277–307.

Dalla Costa, Mariarosa and Selma James. 1972. *The Power of Women and the Subversion of Community*. London: Butler and Tanner.

Davis, Angela. 1981. *Woman, Race & Class*. New York: Random House.

Debord, Guy. 1995. *The Society of the Spectacle*. Translated by Donald Nicholson-Smith. New York: Zone.

Deleuze, Gilles. 1986. *Cinema 2: The Time-Image*. Translated by Hugh Tomlinson and Robert Galeta. Minneapolis: University of Minnesota Press.

Deleuze, Gilles. 1990. *Expressionism in Philosophy: Spinoza*. Translated by Martin Joughin. New York: Zone Books.

Deleuze, Gilles. 1992. "Postscript on the Societies of Control." *October* 59: 3–7.

Deleuze, Gilles and Félix Guattari. 1987. *A Thousand Plateaus: Capitalism and Schizophrenia*. Translated by Brian Massumi. Minneapolis: University of Minnesota Press.

Edwards, Paul N. 1997. *The Closed World: Computers and the Politics of Discourse in Cold War America*. Cambridge, MA: MIT Press.

Ferriera, Francisco H. D. 2021. "Inequality in the Time of Covid-19." *International Monetary Fund: Finance & Development*. https://www.imf.org/external/pubs/ft/fandd/2021/06/inequality-and-covid-19-ferreira.htm.

Foucault, Michel. 1977. *Discipline and Punish: The Birth of the Prison*. Translated by Alan Sheridan. New York: Vintage.

Franklin, Seb. 2021. *The Digitally Disposed: Racial Capitalism and the Informatics of Value*. Minneapolis: University of Minnesota Press.

Galloway, Alexander. 2013. *The Interface Effect*. Malden: Polity.

Grandinetti, Justin. 2022. "'From the Classroom to the Cloud': Zoom and the Platformization of Higher Education." *First Monday* 27, no. 2. https://doi.org/10.5210/fm.v27i2.11655.

Han, Byung-Chul. 2018. *The Expulsion of the Other: Society, Perception and Communication Today*. Malden: Polity.

Harvey, David. 2002. "The Art of Rent: Globalization, Monopoly, and the Commodification of Culture." *Socialist Register* 38: 93–110.

Hudson, Michael. 2021. "Finance Capitalism Versus Industrial Capitalism: The *Rentier* Resurgence and Takeover." *Review of Radical Political Economics* 53, no. 4: 557–73.

Joque, Justin. 2022. *Revolutionary Mathematics: Artificial Intelligence, Statistics and the Logic of Capitalism*. London: Verso.

Katz, Yarden. 2020. *Artificial Whiteness: Politics and Ideology in Artificial Intelligence*. New York: Columbia University Press.

Knight, Frank H. 1921. *Risk, Uncertainty, and Profit*. Boston: Houghton Mifflin.

Kushner, Burton J. 2021. "Eccentric Gaze as a Possible Cause of 'Zoom Fatigue.'" *Journal of Binocular Vision and Ocular Motility* 71, no. 4: 175–80.

Lash, Scott. 2002. *Critique of Information*. London: Sage.

Marazzi, Christian. 2010. *The Violence of Financial Capitalism*. Translated by Kristina Lebedeva. Cambridge: MIT Press.

Massumi, Brian. 2002. *Parables for the Virtual: Movement, Affect, Sensation*. Durham: Duke University Press.

Massumi, Brian. 2021. *Couplets: Travels in Speculative Pragmatism*. Durham: Duke University Press.

Nakamura, Lisa and Peter Chow-White, eds. 2012. *Race After the Internet*. New York: Routledge.

Nakamura, Lisa, Hanah Stiverson, and Kyle Lindsey. 2021. *Racist Zoombombing*. New York: Routledge.

Ngai, Sianne. 2005. *Ugly Feelings*. Cambridge, MA: Harvard University Press.

Noble, Safiya Umoja. 2018. *Algorithms of Oppression: How Search Engines Reinforce Racism*. New York: New York University Press.

Sedwick, Eve Kosofsky. 2003. *Touching Feeling: Affect, Pedagogy, Performativity*. Durham: Duke University Press.

ShonMStudio. 2013. "Apple - IPhone 5 - TV Ad - FaceTime Every Day." *YouTube*. https://www.youtube.com/watch?v=iuFc9jkYmiM.

Skype. 2016. "Skype Group Video Calling for Mobile Phones and Tablets Brings Everyone Together." *YouTube*. https://www.youtube.com/watch?v=7AuhjzxAudY.

Spinoza, Benedictus de. 2018. *Ethics: Proved in Geometrical Order*. Cambridge: Cambridge University Press.

Steve Jobs Videos. 2014. "Steve Jobs: IPhone 4 and FaceTime Introduction - Apple WWDC 2010." *YouTube*. https://www.youtube.com/watch?v=pdkCVoBgHnc&t=5925s.

Sumner, Tyne Daile. 2022. "Zoom Face: Self-Surveillance, Performance, and Display." *Journal of Intercultural Studies* 43, no. 6: 865–79.

Turkle, Sherry 1984. *The Second Self: Computers and the Human Spirit*. Cambridge, MA: MIT Press.

Turkle, Sherry. 1995. *Life on the Screen: Identity in the Age of the Internet*. New York: Simon and Schuster.

Turkle, Sherry. 2011. *Alone Together: Why We Expect More from Technology and Less from Each Other*. New York: Basic Books.

Turkle, Sherry. 2015. *Reclaiming Conversation: The Power of Talk in a Digital Age*. New York: Penguin Press.

Weeks, Kathi. 1998. *Constituting Feminist Subjects*. London: Verso.

Weinberg, Lindsay. 2019. "The Rationalization of Leisure: Marxist Feminism and the Fantasy of Machine Subordination." *Lateral* 8, no. 1. https://doi.org/10.25158/L8.1.3.

Zoom Video Communications. 2021a. "Q4 FY21 Earnings Presentation." https:// investors.zoom.us/static-files/0e5bc6bc-c329-4004-a20b-99b67714e7b8.

Zoom Video Communications. 2021b. "Zoom Video Communications Reports Fourth Quarter and Fiscal Year 2021 Financial Results." *Press Release.* https://investors.zoom .us/news-releases/news-release-details/zoom-video-communications-reports-fourth -quarter-and-fiscal-0.

Zuboff, Shoshana. 2019. *The Age of Surveillance Capitalism: The Fight for a Human Future at the New Frontier of Power.* New York: Public Affairs.

2

Mediating the Death of a Parent

Zoom and the Deathbed Vigil

Susan A. Sci, Regis University

Introduction

*"It was as if the digital amplified the longing for the tactile. This seemed
particularly acute when the need arose to comfort, console, or share grief."*
(Steiner and Veel 2021, *19*)

This is the last picture (Figure 2.1) I have of my mother.[1] She died alone in a
nursing home in Jamaica, Queens, just two days later, on April 22, 2020, due to
complications from Covid-19.

On March 13, 2020, the United States Center for Medicare and Medicaid
Services (CMS) banned all visitors from nursing homes across the country,
impacting over 15,000 facilities. According to Brown (2020), "The blanket ban
[cut] off over 1.5 million older residents from family and friends, with limited
exceptions for end-of-life visits." At my mother's nursing home, all visitors were
completely banned, even for end-of-life-visits.

When my mother began developing Covid-19 symptoms, I knew she would
not survive. Between her dementia, other medical issues, and the virus, my
mother's chances of survival were exceedingly low. Her doctor tried to comfort
me and my siblings, giving us updates every other day and assuring us they were
doing all they could. Nurses reported that she slept most days but, while awake,
was visibly confused and struggling to breathe. Some aides tried to reassure
us there was nothing we could have done if we were there, but this did little to
console us. Not being with my mother through her final days was devastating.
In this distressing situation, the social worker offered what felt like a lifeline—

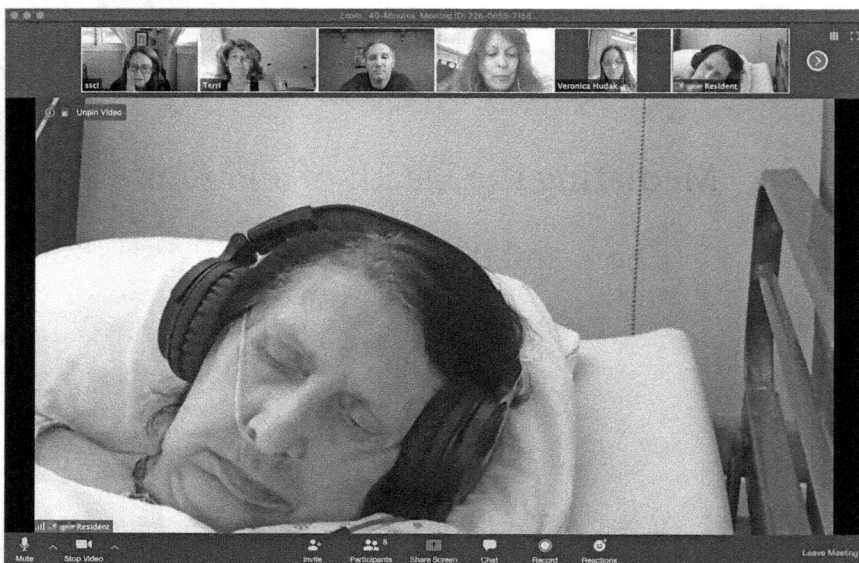

Figure 2.1 Screenshot of a Zoom meeting with my mother, Gloria Sci, and family members. Photograph by Susan A. Sci, April 20, 2020. The name of the nursing home has been removed from the image in order to preserve the anonymity of the facility.

regular Zoom meetings with my mother. She explained that while there wasn't enough staff to stay with my mother throughout the meeting, she could have an iPad set up near the bed and we could, at least, see and speak to my mother. My siblings and I marveled at the opportunities this technology offered us. While I am forever grateful to have seen my mother before her death, having our deathbed vigil mediated by Zoom was far more emotionally and psychologically fraught than I could have imagined. As a media scholar, I couldn't help but wonder why.

The circumstances surrounding my mother's death are not unique. Selman et al. (2021) analyzed experiences like mine via Tweets of bereaved family and friends of Covid-19 victims. The absence of mourning rituals, goodbyes mediated by technology that are appreciated but inadequate, people dying alone at hospitals, hospices, and nursing homes without family and friends present, and the sense of "disrupted bereavement" this creates—these are all part of the Covid-19 pandemic experience.

Overwhelmingly, family members have reported that being unable to visit dying family members at the height of the pandemic was a distressing experience. According to Billingsley (2020), "Facing unprecedented restrictions and challenges, families have experienced unresolved grief when their loved ones have died alone" (276). For many, this has complicated their grieving process (Corpuz 2021; Jordan et al. 2022; Neimeyer and Lee 2022; Nelson-Becker and Victor 2020). Several studies report family members have specifically noted being distraught that "they were unable to be physically close or touch" their dying relatives (Hanna et al. 2021; Selman et al. 2021). In response, healthcare providers turned to numerous communication technologies to help ease this separation.

As Billingsley (2020) notes, "COVID-19 changed the communication landscape in healthcare" (275). Hospitals, nursing homes, and hospice centers turned to videoconferencing platforms such as Zoom, Skype, and Google Meet, as well as video call apps like FaceTime, WhatsApp, and Duo, to bridge the distance between those dying and the families and friends they were leaving behind (Hanna et al. 2021; Schloesser et al. 2021). This move has been lauded by several healthcare researchers as one of the many benefits of these platforms, offering people an opportunity to say their final goodbyes when physical contact was not an option. Some even argue that having loved ones at the bedside, even if just virtually, can help shift an isolated or lonely death, typically considered a "bad death," closer to a "good death" (Parks and Howard 2021). While having the option to use videoconferencing platforms during end-of-life care is widely suggested, most researchers do not recommend using it for patients with cognitive impairments such as dementia, due to the likelihood it would be distressing to the patient and/or their loved ones (Hanna et al. 2021; Schloesser et al. 2021). Other researchers argue that even though videoconferencing is an understandable alternative to in-person visits, it cannot substitute for physical touch and only should be used as a last resort (Schloesser et al. 2021; Selman et al. 2021). While there is no consensus among healthcare researchers on how effective videoconferencing is as a substitute for in-person visits during end-of-life care, there is a consensus about making this technology an available *option* for patients and families and improving on this technology to create a better online experience for both the patient and their loved ones (Billingsley 2020; Elma et al. 2022; Hanna et al. 2021; Moyle et al. 2020; Parks and Howard 2021; Schloesser et al. 2021).

Given the increased role virtual visits will continue to play in end-of-life care, we need to turn a critical eye toward videoconferencing platforms to examine what makes their usefulness as a deathbed technology contested among healthcare researchers. Is there something about the absence of touch and haptics when deathbed visits are mediated through videoconferencing platforms such as Zoom that further complicates this already challenging experience?

To explore this, I will begin by explaining the concept of the "good death" and the elements in which the "goodness" and "badness" of a death are judged. Next, I will focus on deathbed vigils and their importance for both the dying and their loved ones. I will then turn to the concept of mediated death to analyze the difficulties my siblings and I had in adapting our deathbed vigil to Zoom's videoconferencing platform. This will lead to an exploration into the role touch plays during deathbed vigils. I argue that at the deathbed, sole reliance on videoconferencing platforms heightens, rather than dulls, the distance experienced between the patient and loved one due, in part, to its inability to replicate the sensation of touch and the embodied experience of haptics. This limits the possibility for "meaningful communication" that can create a sense of intimacy necessary to reduce the likelihood of complicated grief (Otani et al. 2017). Finally, I will conclude with a discussion of touching vision, a form of tactile-based perception that can help reorient the focus of interaction in mediated deathbed vigils to reinscribe some of the aspects of touch into these experiences.

The "Good Death"

When faced with a life-threatening illness or condition, desires for a "good death" are common. While there is no agreement about what a "good death" is, there are US cultural norms surrounding four aspects of death that impact how people evaluate the seeming "goodness" or "badness" of a death (Campbell 2020). The four aspects of death are: the place of death, one's company in death, the cause of death, and one's manner in facing death (Campbell 2020). For the purposes of this study, I will only focus on the first two aspects—one's place and company at the time of death—since they are directly related to deathbed vigils. Based on these two aspects, I argue that my mother's death could be considered a "good death" given the circumstances of the pandemic. Yet, my siblings and I still felt our grief was complicated, which, I claim, is due to mediating our deathbed vigils through the platform of Zoom.

In the United States, most people believe it is best to die at home, assuming it is a place of comfort, security, and love. As Nelson-Becker and Victor (2020) explain, "the historical image of a good death where one can exercise final control to die at home accompanied by friends and family continues in contemporary times" (2). In Selman et al.'s (2021) study on the Tweets of bereaved family members and friends about their deceased loved ones who died of Covid-19, a common theme was despair over their loved ones dying alone in "alien and overwhelming environments" such as packed hospitals (1271). Similarly, my siblings and I feared my mother would be scared in an unfamiliar and alarming place like a hospital in the midst of New York City's first wave of the Covid-19 pandemic. When faced with the decision to either move my mother to the ICU at the local hospital or keep her at the nursing home she'd been living at for over ten years, we choose the latter. While we could not be at the nursing home or the hospital, we reasoned at least she could be surrounded by care providers she knew and was friendly with.

Keeping my mother at her nursing home was a difficult but ultimately positive decision because it allowed my siblings and I at least some access to her via Zoom. As Selman et al.'s (2021) study showed, family members who were unable to say final goodbyes reported "profound sorrow" and "heartbreak" that many described as "agonizing" and "traumatizing" (1271). Luckily, we were able to have daily Zoom meetings with my mother in the weeks leading up to her death. But like others in this study, "saying goodbye via technology" was "appreciated," yet also "inadequate" with many people describing it as "an unwanted alternative they had to accept" (1271). My siblings and I likewise found it difficult to capture the intimacy we craved at my mother's deathbed over Zoom.

One's company when dying, also known as the deathbed vigil, is another major aspect of evaluating the "goodness" or "badness" of one's death. While understudied, deathbed vigils have a long and culturally varied history.[2] Vigils typically occur when the "terminal stages of dying have commenced" (Kellehear 2013, 110). The tradition of deathbed vigils taps into people's "deep fear of dying alone," due to death's uncertainty and the belief dying alone is a sign of being unloved (Campbell 2020, 611). As Caswell et al. (2022) explain, "The primary reason for family to be at the deathbed is to enact their relationship with the dying person and express the closeness of their bonds" (3). Hence, familial access to the deathbed is widely recognized as integral to compassionate end-of-life care (Downar and Kekewich 2021).

The Deathbed Vigil

The deathbed vigil is the only aspect of a "good death" that has a dual focus on the individual dying as well as the ones left behind, emphasizing the relational quality of both life and death (Caswell et al. 2022). While loved ones try to comfort the one dying, they also comfort each other. According to Kellehear (2013), "Vigils for the dying also play an important role in coming to terms with the prospect of death, to accept the reality of its eventual arrival, for those caring for the dying person" (116). Downar and Kekewich (2021) explain that dying alone, specifically due to the ban on visitors in nursing homes during Covid-19, entails a great deal of psychological risk.

> We know that restrictive visitor policies are associated with a higher frequency of delirium and anxiety in patients. We also know that separation from the patient, the absence of normal death rites, and the disruption of social support networks are risk factors for poor bereavement outcomes. Virtual communication is not feasible for some family members, and might be distressing if the patient is dyspnoeic, delirious, or intubated. (Downar and Kekewich 2021, 336)

Yet, for many, such as my family, there was no other choice; physically visiting loved ones was not an option. During the Covid-19 pandemic, Zoom and other videoconferencing platforms felt like a lifeline in the midst of our impending grief.

This was not our first deathbed vigil; my father died in 2016 after a long illness. My siblings and I were fortunate enough to be with him at his deathbed to say our last goodbyes. While we were not explicitly told what to do during the vigil, the common practices were intuitive. These practices include: holding hands and touching the patient; reminiscing about the past and recounting fond memories; saying final goodbyes and ensuring their legacy; showing affection, emotional support, and gratitude for the impact they have had; making amends for past misdeeds and conflicts; reading and/or singing to the patient; and offering them food and drink. Over Zoom with our mother, my siblings and I tried to enact as many of these practices as we could and included showing old pictures and mementos to help strengthen memories and add emotional depth.

Interestingly, a pattern of our Zoom deathbed vigil seemed to naturally develop. Each day one of us would moderate the call, giving everyone a few minutes to talk directly to our mother. In other words, we managed the calls like business meetings. Although we did not follow an actual agenda, our interactions typically occurred in a noticeable sequence:

Anthony: Tell mom we love her and give examples of her as a great mom.
Terri: Give updates on how the grandkids are.
Debbi: Apologize for not being there and try to explain why.

Michael: Inquire about how mom's feeling and inspire her to fight the virus.
Susan: Remind mom we are here, we love her, we will remember her, and say goodbye.

On good days, she opened her eyes and tried to talk. On bad days, my siblings and I were powerless to help my mother as we witnessed her coughing and struggling to breathe, moaning from aches and pains, and in one particularly harrowing Zoom meeting, watching my mother have a seizure without a doctor or nurse nearby. It was traumatizing to witness her pain in real time and being unable to help her.

Zoom was never intended as a substitute for deathbed vigils, therefore, when planning the platform's user interface, designers were not intending to promote communication based on "closeness" and "intimacy." Rather, Zoom privileges standards of organizational and group communication such as "productivity" and "efficiency." Zoom's platform seemed to discipline our behavior in ways that aligned with organizational communication norms rather than intimacy. Our Zoom deathbed vigils were framed as "meetings" that one "joins"; so, we "joined" our mother's deathbed Zoom "meeting," which positioned us as "participants" struggling to figure out how to meaningfully interact with our mother and each other. Each of my siblings and I attended our vigil "meetings" prepared to comfort our mother and use our limited time wisely. We created an "agenda" of talking points to make sure our mother felt loved and cared for. We shared our screens to show old pictures of our parents as a type of "handout" as we reminisced about the past. We all spoke and listened attentively, doing our best to be respectful and not interrupt each other's time speaking to mom. While our shared purpose, time management, active listening, and agenda helped make our Zoom vigils run smoothly, it also left us with a hollow feeling of longing and a desire for closeness. When not mediated by Zoom, deathbed vigils are often judged on the level of intimacy created in those moments. In the act of mediation, do these deathbed vigil practices still transmit the feelings of love, closeness, and comfort intended? It was impossible to tell based solely on the image of my mom pinned to my screen (Figure 2.2). It quickly became clear that mediating the deathbed vigil experience was more complicated than we had hoped.

Mediated Death

A mediated death is, first and foremost, a representation of a death. As a representation, individuals can virtually or remotely experience another's death

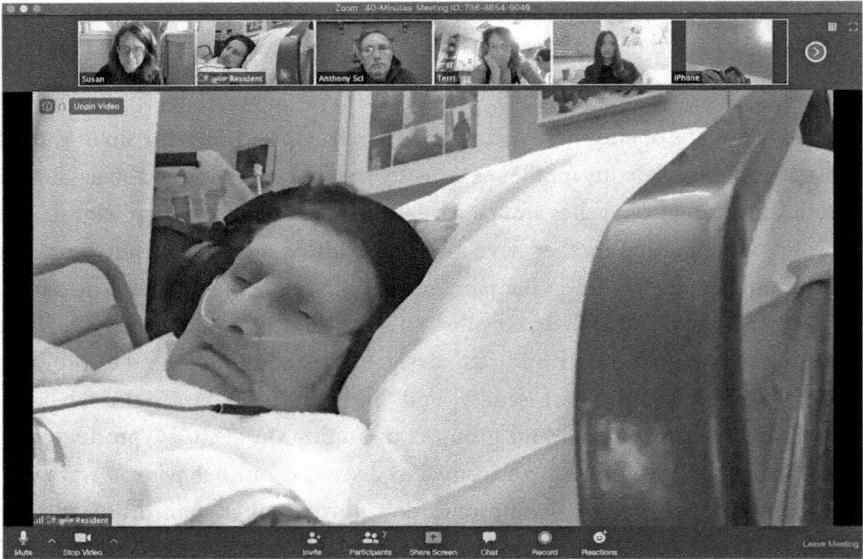

Figure 2.2 Screenshot of my Zoom meeting with my mother, Gloria Sci, and family members. Photograph by Susan A. Sci, April 12, 2020. The name of the nursing home has been removed from the image in order to preserve the anonymity of the facility.

without being physically present. When a death is mediated, our knowledge of a person's passing is derived from news stories, TV reports, live streamed accounts, social media posts, and so on. Therefore, a mediated death can be "re-lived, experienced, and shared regardless of era or location, so long as there is internet access" (Sumiala 2022, 41). Early research on mediated death focused on how the constraints of mass media limited which deaths were shared publicly. Consequently, researchers found that for a death to be covered by mass media it had to occur publicly (i.e., executions, crimes, tragedies), and it had to be exceptional due to either the circumstances, the manner in which it occurred, or the identity of the deceased. These mass-mediated deaths fell into the category of newsworthy or grievable deaths, and journalists were the gatekeepers of which deaths were considered important enough to be reported on.[3] However, with the advent of digital and social media, one's death no longer had to be public or exceptional to be mediated. People share the news of a loved one's passing widely on all social media platforms including Facebook, Instagram, Twitter, and TikTok. On the internet, virtual memorials and tributes have become commonplace. And now—post-lockdown—several funeral parlors still offer remote and streaming funeral options through platforms like

Zoom and TribuCast, as either a compliment to or substitute for in-person ceremonies. As media and technology have evolved, so have our representations of death from mostly mass-mediated spectacles to hybrid, interactive, mediated experiences.

Due to my mother's dementia and her Covid-19 symptoms, our Zoom deathbed vigils fell somewhere between these two types of mediated deaths—a hybrid, interactive experience and a spectacle. Since Zoom is a videoconferencing platform, the representation of the deathbed vigil is both synchronous and interactive, creating temporal copresence and coexistence within a shared virtual space. This enables participants to experience a basic form of "telepresence." According to Steuer (1992), "telepresence is the extent to which one feels present in the mediated environment, rather than in the immediate physical environment" (6), a sensation of immersion that is intensified via increased interactivity and sensory vividness. However, in my situation, as a representation of my mother's deathbed, the Zoom vigils were a constant reminder of my physical absence. There was no sense of immersion; rather the lack of our physical presence was always salient, turning what should have been private, intimate moments into mediated spectacles of my mother's suffering that my siblings and I watched together. After sitting at my father's deathbed years earlier, my siblings and I knew what to expect; however, navigating it over Zoom posed several problems. The most significant issue was how to get my barely conscious mother to understand what Zoom was and interact with us to create a more immersive telepresence with her. We could not touch her to alert her to our presence, and we could not whisper in her ear to gently rouse her. Without her interaction, we would become mere passive spectators to her death.

Prior to digital and social media, telepresence at a mediated death was impossible. Bell (1997), in her study on live televised funerals, argued that these funerals' personal focus and synchronous, live broadcast created an "intimate distance" between the viewers and the subject. According to Chouliaraki (2006), intimate distance entails two dimensions of the spectator-sufferer relationship: proximity-distance and watching-acting. As she explains:

> [The spectator-sufferer relationship] expresses the tension between proximity and distance—feeling close but unable to approach the person—and between watching and acting – seeing the [person] on the screen, but being unable to do something with or for that person. (21)

The broadcast of televised death rituals, such as funerals, positions viewers as intimate spectators whose main role in the mediated death is to watch. Consequently, mass-mediated representations of death were often spectacles of distant suffering and invited spectators to feel pity rather than a "natural sentiment of love and care" (Chouliaraki 2006, 11).

The concept of intimate distance and its paradoxical dimensions aptly describes the experience of participating in my mother's Zoom deathbed vigil. Over two weeks, at the same time daily, my siblings and I "tuned into" my mother's deathbed. However, it was during our first "meeting" that our intimate distance was the farthest and our positions of spectator-sufferer were the most salient. We were Zoom novices but excited about the prospect of seeing our mother.

"Hey, look, there's mom!"
"Mom, can you hear me?"
"Mom, it's me, Susan."

Statements like this gave way to questions about how she was. "Mom, you look pretty good, how do you feel?" Then, the slow realization that my mother had no idea what videoconferencing was, nor was she about to figure it out in the throes of Covid-19 and dementia. Before our eyes, the hope of interacting with my mother quickly dissolved, which, in turn, seemed to lengthen our physical, mental, and emotional distance.

"Do you think she knows it's us?"
"Mom, can you open your eyes? It's us, your kids."
"Does she think we're a TV show?"

As a videoconferencing platform, Zoom's interface is preset to maximize the visual display of speaking participants. Unless changed, the screen can become a blur of quick cuts between participants vying for a chance to talk. Frantically trying to assess how ill our mother was, the first meeting was a postmodern nightmare of worried faces and a cacophony of concerned voices. My mother stared blankly at us. Her fatigue and confusion were palpable; she closed her eyes. I pinned my mother's screen and we fell silent just watching my mother in her bed.

The call ended with a deep sense of sorrow and despair as well as pity for my mother. She was trapped there alone in her nursing home without any of us being next to her to help comfort her. Did my mom, who had never seen Zoom before, think this was a hallucination? Images of us blurring together, talking over each other, yet not one of us sitting on her bedside—how could my mother make sense of this? The intensity of my desire to comfort my mother could not be translated through Zoom. Rather, it came across as frantic images calling to her like a bad 1990s commercial. In her frail state, the only response she could give was to tune us out as a way to find calmness, turning our potential interaction into spectatorship. My siblings and I watched my mother and commented between ourselves about how she seemed, how bad her situation was, and how we were coping with the pandemic. While my siblings and I had a stronger sense of telepresence between us, without her active participation the pinned image of my mother became our shared spectacle of pity.

In order to have our Zoom deathbed vigils not remain a spectacle of pity, my siblings and I had to actively work to find what we could do to engage my mother. We tried all the norms of deathbed vigils we could over Zoom—affirm our love for her, reminisce about good times, show gratitude for all she did for us, make amends for past misdeeds—all to no avail. We found ourselves actively having to remind each other that this was happening in *real time*; our mother was, indeed, dying from Covid-19, and we had to make this time meaningful no matter how difficult it was to accept and how surreal it felt. These Zoom "meetings" were not spectacles of suffering; we were witnessing our mother slowly die from Covid-19 right before our eyes but completely out of our grasp. Our grief was mediated in this liminal space of proximity and distance by Zoom, where we could watch but not act.

A mediated witness is one who sees a direct representation of another's suffering and/or death unfold. Watching live war coverage, breaking news of a mass shooting, or a person vlogging about their cancer struggles are examples of mediated witnessing. "As a moral obligation, witnessing suffering and death always calls for action, such as taking part in public mourning, grief, and commemoration" (Sumiala 2022, 107). Calls to action are not just about catharsis through grief but also about the need for political and social action as a form of responsibility based within the conditions of the victim's suffering and death. But what was my family's call to action? In my mother's case, her death was inevitable, but her suffering alone was not. Consequently, our moral imperative was tied to her isolation. As our grief was mediated by Zoom, my siblings and I needed to comfort her and express our love to help mitigate *our* guilt and despair about not physically being there. *But what did my mother need?* What was it that touch would communicate to my mother on her deathbed that felt so intensely absent over Zoom?

The Significance of Touch at Deathbed Vigils

Touch is frequently identified as an important form of nonverbal communication at a loved one's deathbed (Khalid 2020; Selman et al. 2021; Woodhouse 2004). Presence is often conflated with touch and understood as a means of providing comfort and showing affection, especially when the patient is unconscious. Woodhouse (2004) observed that touching at the deathbed took many forms, such as: kissing the dying's cheek, holding their hand, caressing their shoulder, leg, or back. Especially upon entry, visitors were drawn to their loved one, reaching out for initial contact. "Reaching out to touch or caress an animate object," such as a loved one, explains Paterson (2007), and "the immediacy of sensation is affirmatory and comforting, involving a mutual co-implication of one's own body and another's presence" (3). Touching is an expression of proximity and immediacy in which we engage the materiality of our world and those in it.

To be close enough to touch is to experience a "sensuality of the flesh, an exchange of warmth, a feeling of pressure, of presence, a proximity of otherness that brings the other nearly as close as oneself. Perhaps closer" (Barad 2012, 206). The sense of touch ignites both external cutaneous sensations via the bidirectional experience of simultaneously touching and being touched and the internal sensations of bodily position through muscle tension (proprioception), movement of the body and limbs (kinesthesia), and balance in relation to what we are touching (vestibular). Combined, this makes up our haptic system. While our skin, as an organ of touch, enables us to identify texture, temperature, pressure, and pain, it does not capture the full embodied experience of what we feel when we are touching and being touched. How our body is oriented toward the person/object we are touching, the fluidity of the movement of the touch and how our bodies respond to it, how our balance adjusts to account for the (in)stability of what we are touching—all of this impacts how touch feels to us and those we are touching. Khalid (2020) explains that "the practice of presence [through touch and silence] further points towards the centrality of embodiment: to truly 'be with' another is to be present bodily, as one shares time and space with the dying person. Palliative care thus emphasizes the significance of dying as an *embodied* experience" (154). Combined, this makes touch at one's deathbed a complex and meaningful experience.

To touch is to be attentive to the other as "other": appreciating the other's alterity while simultaneously recognizing the effects it has on oneself via our haptic sensations. In her theological analysis of touch and silence at the deathbed, Khalid (2020) asserts: "In the case of touch, the creative affirmation of the other becomes a concrete recognition of the other's *particularity*, inasmuch as touch constitutes and sustains us as flesh—as the distinctive, embodied person that we are" (159). Over Zoom, my mother's name was never listed on her image; rather, she was just identified as "resident" of her facility.[4] In the final days of her life, my mother's identity, her particularity as "Gloria Sci," was literally overwritten by her institutional role. "Gloria" did not exist in our deathbed "meetings," only "resident" did. By framing each participant within the same virtual confines of a visual image identified by name, we were represented in an equitable manner but stripped of our particularity. To use Sherry Turkle's phrase, we were "alone together" trying to feel a sense of embodied copresence with only audiovisual technology available.

Relationality is central to our experience of touch. Our skin is often thought of as our bodily boundary, hence our tactile feelings "provide a sense of the limits of the body" (Ratcliffe 2008, 306). Both the external and internal sensations of touch are impacted by the connection, or contact, that occurs between the toucher and the touched. As Bellacasa (2017) explains:

> Understanding contact as touch intensifies a sense of the co-transformative, in the flesh effects of connections between beings. Significantly, in its quasi-automatic evocation of close relationality, touching is also called upon as the experience par excellence where boundaries between self and other are blurred. (96)

While touch acknowledges the alterity and particularity of the other, it can also blur the lines between the self and other, especially within the movement of the caress. The caress makes the separation between subject and object barely discernible, folding into the other and vice versa. As a form of embodied enmeshment, the caress is a moment of intercorporeality par excellence, "the other's body speaks directly to my body, which responds directly" (Dolezal 2021). Due to the intercorporeality of touch, Paterson (2007) argues that it can "cement an empathic and affective bond, opening an entirely new channel of communication" (3). This channel blurs the boundaries between self and other, creating not just a feeling of empathy but also a sensation of feeling-with that person.

In the case of palliative care, Khalid (2020) argues that being at the deathbed is a way of "accompanying a dying person with physical presence and a gentle touch . . . [so] the dying person can recognize that he or she is loved and held in memory" (158). Every day my siblings and I virtually sat by my mother's bedside, recounting fond memories for her, showing her old pictures, and expressing how much we loved and cherished her. Ending every call with "Bye Mom. We love you," we wondered if she would live through the night. But we weren't on her bed. Looking at and hearing my mother every day for two weeks over Zoom, knowing I could never feel her again, was, without hyperbole, traumatic. Relying on audiovisual technology to replicate the comfort of tactile proximity on the cusp of death is an impossible dilemma. I longed to comb her hair, to hold her hand, to cover her perpetually chilly feet in her favorite blanket, to feel her breath, to hand her water. Most, if not all, communication technology was not designed to mediate the experience of death between loved ones. In my mother's disoriented state brought about by dementia and Covid-19, the sensate experience of sight and sound Zoom relies on could be confusing, but it was all we had. I empathized with my mother and wanted desperately to ease her fears and pain, to bring her a sensation of comfort and care to what I imagined was a terrifying experience, in part because it *appeared* terrifying, both in the media and over Zoom. But *how did my mother long to be touched?* And if we did touch her, *what would we feel-with her?* Could I adapt to the constraints of videoconferencing technology to create an approximate form of touch to achieve this? If so, how?

Touching Vision

My siblings and I had to accept that touch and the experience of intercorporeality that can occur with it was impossible to replicate over Zoom. Our telepresence had to substitute for our physical presence, but it could not recreate it. The lack of intercorporeality and the tacit, unspoken understanding this bodily communication creates is at the core of my complicated grief. My siblings and I searched for the right words, pictures, and stories to share with my mother to bring her comfort and express our love. But we had to slowly embrace the realization that the way touch communicates empathy and affect is unparalleled. As the days ticked by, my siblings and I began focusing on what would truly touch my mother, make her feel recognized as distinctly herself, blur the boundaries between us, and feel-with her without the luxury of touch.

Instead of striving for an unattainable intercorporeality, what made me closer to my mother was attempting to combine touching with seeing, haptic with optic, in what Bellacasa (2017) calls "touching visions." As she explains, "What these visions that play with vision as touch and touch as vision invite to think is a world constantly done and undone through encounters that accentuate both the attraction of closeness as well as awareness of alterity . . . [that] require a situated ethicality" (115). Touching visions account for both the proximity of haptics and the distance of optics. In other words, I had to bridge the feeling of closeness and familiarity with who I knew my mother was with an appreciation of her "otherness" outside of the bonds of our immediate family. I had to let go of the "figure," or vision, of my mother as "parent" as well as the "possession," or hold, her presence had in my life and vice versa. I had to begin by asking myself: *who and what do I touch when I touch my mother?*

The day before my mother died, I wandered around my house looking for something to show or read to bring her comfort. It was clear she was getting worse and her time was limited. I noticed the book my maternal grandmother wrote about our family and immediately knew I needed to read it to her. Throughout her life, my mother, Gloria, felt an immense bond with her own mother, Lena. More than anything, Gloria longed to feel as if Lena was proud

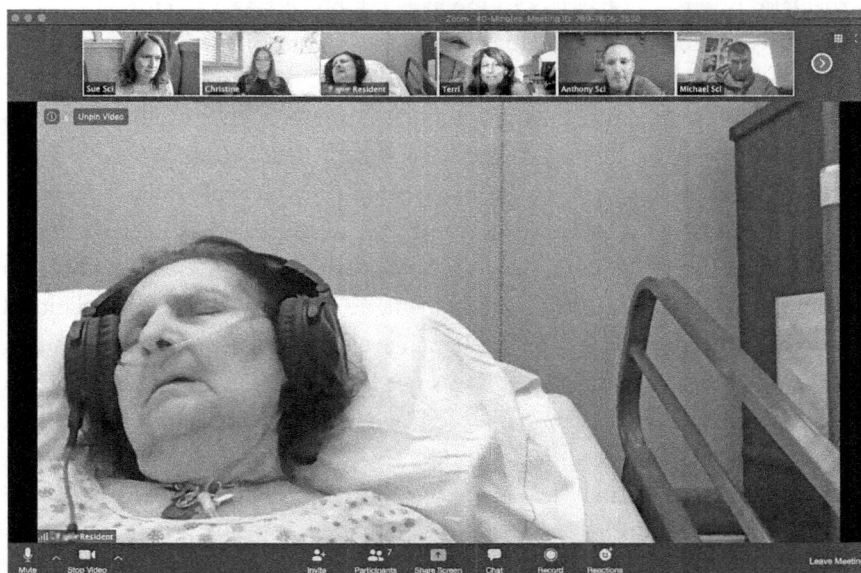

Figure 2.3 Screenshot of my Zoom meeting with my mother, Gloria Sci, and family members. Photograph by Susan A. Sci, April 18, 2020.

of her. As the oldest child of immigrant parents, Gloria tried to be the dutiful daughter but often questioned if it was enough to make her mother proud. On her deathbed, what Gloria needed to feel was reassurance that she was, indeed, enough: a sentiment only her own mother could assure her of. The trace of my grandmother's pride was etched into the story she wrote about my parents' wedding day. Thankfully, my mother made it through the night, and I read it to her over Zoom the next day. Two hours later I got the call that my mother died. For all the trauma I felt over Zoom, I'm eternally grateful I had this one moment with her. The intimacy I felt with my mother and siblings as I read from her mother's book was the closest I felt to touching her (Figure 2.3).

Conclusion

Mediating the death of my mother over Zoom was one of the most profound experiences of my adult life and unique to other Covid-19 Zoom situations. Simultaneously overjoyed to see and hear my mother and overwhelmed by my inability to touch her, my mother's Zoom deathbed vigil challenged my expectations, creating barriers to the grieving process and a sense of closure. While I will be forever grateful to have had the opportunity to be at my mother's virtual deathbed, it was still a fraught experience.

While videoconferencing platforms like Zoom can give families access to remotely visit loved ones, when used as a substitute for deathbed vigils it can complicate the grieving process due to its sole reliance on audiovisual technology in lieu of haptics and touch. To ease the lack of touch, loved ones should be prepared to not only focus on their personal relationship with the dying but also recognize that individual as a singular, unique person beyond their relational ties to help generate the sensation of touch even when many miles away.

Even though my mother's death was not ideal, it still could be considered a "good death." She died in a familiar place. Her death wasn't arbitrary, meaningless, or embarrassing. She faced death in the same way she did life, with some fear and anxiety but ultimately with a sense of resolve. And, ultimately, her loved ones were at her deathbed, just not in the manner any of us expected or fully understood. This is just the beginning of mediated deathbed visits and the more we know about how mediated deathbed vigils can impact those going through it, the greater the urgency to address this issue.

Notes

1 My siblings, Anthony Sci, Theresa Ezratty, Debra Hudak, and Michael Sci, as well as my nieces, Veronica Hudak and Christine Sci, have consented to having their image used in this chapter. Additionally, my siblings have given their consent for their comments to be represented for the academic purposes of this chapter.
2 Kellhear (2013) explains that the intimate nature of deathbed vigils is the reason they are not widely studied. Delving into others' experiences of these rituals is considered too intrusive.
3 For a detailed literature review on mass-mediated death, see Sumiala (2022).
4 Note that the images included in this chapter have been altered to remove the name of my mother's nursing home to maintain the anonymity of her facility.

References

Barad, Karen. 2012. "On Touching – The Inhuman That Therefore I Am." *Differences* 23 (3): 206–23. https://doi.org/10.1215/10407391–1892943.

Bell, Catherine. 1997. *Ritual: Perspectives and Dimensions.* New York: Oxford University Press.

de la Bellacasa, Maria Puig. 2017. *Matters of Care: Speculative Ethics in More than Human Worlds. Matters of Care.* Minneapolis: University of Minnesota Press.

Billingsley, Luanne. 2020. "Using Video Conferencing Applications to Share the Death Experience during the COVID-19 Pandemic." *Journal of Radiology Nursing* 39 (4): 275–7. https://doi.org/10.1016/j.jradnu.2020.08.001.

Brown, Bethany. 2020. "US Nursing Home Visitor Ban Isolates Seniors." Human Rights Watch, March 20, 2020. https://www.hrw.org/news/2020/03/20/us-nursing-home -visitor-ban-isolates-seniors.

Campbell, Stephen M. 2020. "Well-Being and the Good Death." *Ethical Theory and Moral Practice* 23 (3/4): 607–23. https://doi.org/10.1007/s10677-020-10101-3.

Caswell, Glenys, Eleanor Wilson, Nicola Turner, and Kristian Pollock. 2022. "'It's Not Like in the Films': Bereaved People's Experiences of the Deathbed Vigil." In *OMEGA-Journal of Death and Dying*, 1–18. Sage OnlineFirst, October 14, 2022. https://doi.org/10.1177/00302228221133413.

Chouliaraki, Lilie. 2006. *The Spectatorship of Suffering.* London: Sage Publications.

Corpuz, Jeff Clyde G. 2021. "Beyond Death and Afterlife: The Complicated Process of Grief in the Time of COVID-19." *Journal of Public Health (Oxford, England)* 43 (2): e281–2. https://doi.org/ 10.1093/pubmed/fdaa247.

Dolezal, Luna. 2021. "Intercorporeality and Social Distancing." *The Philosopher,* November 18, 2021. https://www.thephilosopher1923.org/post/intercorporeality -and-social-distancing.

Downar, James and Mike Kekewich. 2021. "Improving Family Access to Dying Patients during the COVID-19 Pandemic." *The Lancet Respiratory Medicine* 9 (4): 335–7. https://doi.org/10.1016/s2213-2600(21)00025-4.

Elma, Asiana, Deborah Cook, Michelle Howard, Alyson Takaoka, Neala Hoad, Marilyn Swinton, France Clarke, Jill Rudkowski, Anne Boyle, Brittany Dennis, Daniel Brandt Vegas, and Meredith Vanstone. 2022. "Use of Video Technology in End-of-Life Care for Hospitalized Patients during the COVID-19 Pandemic." *American Journal of Critical Care: An Official Publication, American Association of Critical-Care Nurses* 31 (3): 240–8. https://doi.org/10.4037/ajcc2022722.

Hanna, Jeffrey R, Elizabeth Rapa, Louise J Dalton, Rosemary Hughes, Tamsin McGlinchey, Kate M Bennett, Warren J Donnellan, Stephen R Mason, and Catriona R Mayland. 2021. "A Qualitative Study of Bereaved Relatives' End of Life Experiences during the COVID-19 Pandemic." *Palliative Medicine* 35 (5): 843–51. https://doi.org/10.1177/02692163211004210.

Jordan, Timothy R., Amy J. Wotring, Colette A. McAfee, Derek Cegelka, Victoria R. Wagner-Greene, Mounika Polavarapu, and Zena Hamdan. 2022. "The COVID-19 Pandemic Has Changed Dying and Grief: Will There Be a Surge of Complicated Grief?" *Death Studies* 46 (1): 84–90. https://doi.org/10.1080/07481187.2021.1929571.

Kellehear, Allan. 2013. "Vigils for the Dying: Origin and Functions of a Persistent Tradition." *Illness, Crisis & Loss* 21 (2): 109–24. https://doi.org/10.2190/il.21.2.c.

Khalid, Hina. 2020. "At the Bedside: A Theological Consideration of the Role of Silence and Touch in the Accompaniment of the Dying." *Scottish Journal of Theology* 73 (2): 150–9. https://doi.org/10.1017/s0036930620000277.

Moyle, Wendy, Cindy Jones, Jenny Murfield, and Fangli Liu. 2020. "'For Me at 90, It's Going to Be Difficult': Feasibility of Using iPad Video-Conferencing with Older Adults in Long-Term Aged Care." *Aging & Mental Health* 24 (2): 349–52. https://doi.org/10.1080/13607863.2018.1525605.

Neimeyer, Robert A. and Sherman A. Lee. 2022. "Circumstances of the Death and Associated Risk Factors for Severity and Impairment of COVID-19 Grief." *Death Studies* 46 (1): 34–42. https://doi.org/10.1080/07481187.2021.1896459.

Nelson-Becker, Holly and Christina Victor. 2020. "Dying Alone and Lonely Dying: Media Discourse and Pandemic Conditions." *Journal of Aging Studies* 55 (December): 100878. https://doi.org/10.1016/j.jaging.2020.100878.

Otani, Hiroyuki, Saran Yoshida, Tatsuya Morita, Maho Aoyama, Yoshiyuki Kizawa, Yasuo Shima, Satoru Tsuneto, and Mitsunori Miyashita. 2017. "Meaningful Communication before Death, but Not Present at the Time of Death Itself, Is Associated with Better Outcomes on Measures of Depression and Complicated Grief among Bereaved Family Members of Cancer Patients." *Journal of Pain and Symptom Management* 54 (3): 273–9. https://doi.org/10.1016/j.jpainsymman.2017.07.010.

Parks, Jennifer A. and Maria Howard. 2021. "Dying Well in Nursing Homes during COVID-19 and Beyond: The Need for a Relational and Familial Ethic." *Bioethics* 35 (6): 589–95. https://doi.org/10.1111/bioe.12881.

Paterson, Mark. 2007. *The Senses of Touch: Haptics, Affects and Technologies.* Oxford: Berg Publishers.

Ratcliffe, Matthew. 2008. "Touch and Situatedness." *International Journal of Philosophical Studies* 16 (3): 299–322. https://doi.org/10.1080/09672550802110827.

Schloesser, Karlotta, Steffen T Simon, Berenike Pauli, Raymond Voltz, Norma Jung, Charlotte Leisse, and Agnes van der Heide, et al. 2021. "'Saying Goodbye All Alone with No Close Support Was Difficult'- Dying during the COVID-19 Pandemic: An Online Survey among Bereaved Relatives about End-of-Life Care for Patients with or without SARS-CoV2 Infection." *BMC Health Services Research* 21 (1): 998. https://doi.org/10.1186/s12913-021-06987-z.

Selman, Lucy E, Charlotte Chamberlain, Ryann Sowden, Davina Chao, Daniel Selman, Mark Taubert, and Philip Braude. 2021. "Sadness, Despair and Anger When a Patient Dies Alone from COVID-19: A Thematic Content Analysis of Twitter Data from Bereaved Family Members and Friends." *Palliative Medicine* 35 (7): 1267–76. https://doi.org/10.1177/02692163211017026.

Steiner, Henriette and Kristin Veel. 2021. *Touch in the Time of Corona: Reflections on Love, Care, and Vulnerability in the Pandemic.* Berlin: De Gruyter.

Steuer, Jonathan. 1992. "Defining Virtual Reality: Dimensions Determining Telepresence." *Journal of Communication* 42 (4): 73–93. https://doi.org/10.1111/j.1460-2466.1992.tb00812.x.

Sumiala, Johanna. 2022. *Mediated Death.* Medford, MA: Polity Press.

Woodhouse, Jan. 2004. "A Personal Reflection on Sitting at the Bedside of a Dying Loved One: The Vigil." *International Journal of Palliative Nursing* 10 (11): 537–41. https://doi.org/10.12968/ijpn.2004.10.11.17133.

Zoom Etiquette Guides

Negotiating between Workplace Professionalism and Gendered Homeplace Surveillance in the Videoconferencing Borderlands

Jacquelyne Thoni Howard, Tulane University

In borderland spaces, different social and cultural mores collide along a boundary and manifest new ways of existence (DuVal 2010). Bureaucrats working on behalf of a prevailing entity, such as an imperial metropole, attempt to enforce dominance over other groups who are sustaining their own customs. In borderlands spaces, individuals often blend the traditions of both divides to create new cultural norms and identities within that space situated between distinctions (Hämäläinen and Truett 2011; Perales 2013, 165, 167). Historians often study these power dynamics by observing the relationship between people acting from the "cores and peripheries" of power (DuVal 2010). Borderlands spaces may occur along physical borders but can also embody metaphorical, cultural, theoretical, and organic expanses. Work and home represent two separate locations that both have their own rules of engagement. During the Covid-19 pandemic, borderland spaces, situated between home and work, emerged through videoconferencing technology, such as Zoom, which allowed employees to complete work meetings from home, while at the same time permitting personal life to transcend the boundaries of work.

Borderlands can form in different types of spaces, ranging from colonial territories to virtual environments. Despite different modalities, similarities in policies, enforcement, and resistance often emerge across different borderlands. Historians of North and South American colonial encounters illustrate that within borderland spaces, people of different backgrounds purposely come together to exchange ideas and goods and in the process create new cultural identities

(Metcalf 2006, 1–16). In some cases, people known as "go-betweens," historically referred to as mixed-heritage people of Indigenous and European descent across North and South American colonies, gain skills to perform within the bounds of both cultures (Metcalf 2006, 1–16). Borderland spaces may also extend beyond colonial settings and into digital spaces such as social media platforms and video games where people build cyber communities, develop personas, or exchange digital currency (Steele 2021, 41–65; Fornäs 2002, 1–45). During the lockdown period, while employees attended work meetings from their homes using videoconferencing software, employees inhabited a virtual borderland space situated between the mores of their homes and their professional work cultures. In these Zoom borderlands, inhabitants could not always compartmentalize their work personas and their home identities. Instead, users often combined workplace and homeplace behaviors to create new virtual work standards. Many employees, such as parents and caretakers, also acted as a type of "go-between," constantly renegotiating the priorities of home and work throughout a workday, as they straddled both their work and home personas often at the same time (Metcalf 2006, 1–16).

Within borderlands, people who benefit from the dominant culture often view new behaviors as threats to their way of life. Authority figures often use surveillance to guard against the mixing of customs around borders. As a response, people in power often use deputies of their system within the borderlands, such as colonial administrators, online moderators, and supervisors, to police occupants. This regulation usually takes the form of official policies and evaluations of social virtue (Muñiz 2022, 19–38; Perales 2013, 167). In digital spaces, institutions use digital surveillance applications, such as data analytics and metrics, to entrench their power by monitoring the behavior of users (Blum 2022). In March 2020, as lockdown protocols moved much of professional work into the home, supervisors attempted to control the behavior of their workers and their families within their personal space.

During this period, business advocates reapplied surveillance mechanisms created during the rise of capitalist corporations in the Progressive era that connected employees' personal identities and homeplace behaviors to workplace productivity. Since then, corporations have used multiple forms of surveillance to monitor employees at home and at work to increase profits. As Michel Foucault (1995) indicates, surveillance achieves a form of "discipline" that depends on multimodal types of observation. In the workplace, these different observation points include: employers watching their employees, lateral enforcement among work peers, and employee self-monitoring. During the Progressive era, employers such as Ford Motor Company used top-down surveillance to seemingly gauge the productivity and morality of their workers. They coded language and body standards as metrics of professionalism (Foster 2000, 2–6; Gray 2019; The Henry Ford n.d.). Employers, however, shaped these professional behaviors around social biases involving class, race, and gender norms that entrenched systemic discrimination (Foster 2000, 2–6; Gray 2019). Ford Motor Company's Sociological Department's surveillance even reached into employees' homes as they "[taught] wives about home care, cooking, and hygiene," to support men's ability to produce at work (The Henry Ford n.d.).

During the Progressive era, social commentators also produced etiquette guides to influence mores around work and home through lateral and self-monitoring surveillance. Etiquette guides require the critic to take on an "authority on behavior" stance to protect the dominant social class over those they perceive as social others (Foster 2000, 3, 6). By observing etiquette, especially in the homeplace, employers and some peers thought that they could measure workplace productivity by looking for signs of professionalism in the home while encouraging the erasure of personal social identities around race, class, and gender as a form of moral discipline (Foster 2000, 2, 10). Even in 2020, employers were still attempting to control the private lives of their employees through etiquette guides that marked standards of professionalism around clothes and beauty standards.

While employers have a long history of using surveillance at work and home to bolster production, the situation that occurred in March 2020, requiring employees to work from home, initiated new pathways for employers to surveil domestic life. For employers, videoconferencing technologies served as a panopticon due to the self-monitoring nature of seeing oneself as seen by observers who may or may not be looking at certain workers (Emigh and O'Malley 2020). In prisons, Foucault (1995) explained that the panopticon functioned:

> to induce . . . a state of conscious and permanent visibility that assures the automatic functioning of power . . . the inmates should be caught up in a power situation of which they are themselves the bearers. . . . Visible: the inmate will constantly have before his eyes the tall outline of the central tower from which he is spied upon. Unverifiable: the inmate must never know whether he is being looked at any one moment; but he must be sure that he may always be so. (201)

Like the prison system, workplaces used surveillance to induce a fear of being watched to subconsciously regulate the performance of their employees (Brivot and Gendron 2011, 140; Foucault 1995, 228). In the Zoom borderlands, however, even though observers possessed a limited view or "partial perspective," employers and peers held authority over those they watched (Taurek 2019, 29; Haraway 1988, 575–99). Starting in March 2020, etiquette writers encouraged users to measure professionalism and workplace morality standards among their employees, peers, and themselves based on what they could see via Zoom (Andersen 2021; Montgomery 2019; Weidinger 2020). Critics used articles and social media posts to offer advice about the proper and improper way to present their professional identities from home. This guidance illustrated the many ways that employers often made direct and indirect commentary on what private and therefore gendered behaviors they viewed as unacceptable in the new home-work borderlands.

Expectations in the Zoom Borderlands

As workers moved their workstations into their homes during the lockdown period, many employers expected their control over employees' productivity to extend into their employees' homes. This desire to control employees' behavior during a disaster also reached into the physical space of the home and included those in the home who did not work for the company (Blum 2022). Employees monitored employees' performance using Zoom via the scene made available through a camera, even if it represented a fleeting moment of order while on a video call. Despite requests for employees to maintain the boundaries of home and work within the videoconferencing screen, many workers could not always meet these expectations. This burden mostly fell on women and underrepresented groups because of their roles as caretakers, home and family size, and diverse cultural environments. These expectations point to the highly gendered but also economic and racial inequalities that still inform practices in white-collared workspaces that privilege those who can mimic workplace behaviors designed by men. In the spaces outside of (and sometimes within) the digital box, employees shared space with their families also facing the crisis. These members, often mix generation, included other working people, children attending virtual school or without usual childcare arrangements, elderly individuals, and pets that normally occupied this space during the day (Bender 2020; Weik 2021; Wik 2021).

Business advocates extended advice on behaviors often involving gender and family that they deemed either proper or inappropriate in the Zoom borderlands through online etiquette guides. Some of these manuals illustrated ways that supervisors attempted to control the social and cultural behaviors of their workers and their families within their homes. Within these Zoom borderlands, managers and coworkers made judgments about employees' work ethic by pointing to the social and cultural aspects of the home that represented qualities socially deemed as personal and therefore feminine. For example, these guides questioned if the background as seen through the video box looked like a professional workplace or a home (Emily Post 2020; Zight 2020). They suggested workers physically separate themselves from the presence of children and pets to give the perception that workers were not distracted by the feminized duties of caretaking (Bender 2020; Maddox 2020). They judged the clothing typically worn in the home by employees as being too casual for conducting serious work (Andersen 2021). They also used beauty standards that subjugate women, especially women of color, to control how employees represented themselves on camera (Buchanan 2020; Muller 2020).

Often, these observers indicated what they perceived as the "stakes" of not complying with new professional standards when using videoconferencing software. As one commentator explained:

> If you have a Zoom meeting on the calendar, or think you might in the future, you'd be wise to avoid making a fool of yourself and maintain a professional persona while video conferencing. Utilizing proper etiquette in these meetings will not only keep you on the boss's good side but it will help ensure you don't become the subject of your own humiliating viral video. (Weidinger 2020)

In addition to manager oversight, employees also had to worry about peers publicly posting evidence of social mishaps on social media (Dray 2020). While warning readers that mistakes could lead to shame, they also pointed to examples of workers behaving inappropriately. Unsuitable behavior discussed by commentators, however, ranged significantly from inappropriate sexual misconduct to a family member disrupting a meeting, to having private conversations while not on mute (Safronova 2020). The *New York Times* recounted that:

> Many of us are now living with grown-up versions of the "I came to school naked" nightmare . . . The nude self-portrait you painted popping up in a video chat on the wall behind you. Your collection of cannabis cookbooks appearing in the background of a video call with your boss. (Safronova 2020)

During an unprecedented deadly pandemic, business advocates used their platforms not to encourage well-being but instead to warn employees about the stakes of maintaining professionalism.

During the lockdown period, many Zoom etiquette guides devoted copious attention to the need for employees to control personal spaces for work. Commentators encouraged workers to regulate the contents that made up the background of their video box and sound while on Zoom. Critics called for employees to "stage the video properly: make sure your background is business appropriate" and to keep their work areas organized (Emily Post 2020; Zight 2020). When employees could not create a sterile setting in which to conduct their meetings, devoid of home materials, business advice columnists instructed them to use Zoom's virtual background feature, where users could upload a virtual wallpaper that obscured the space behind where they positioned themselves and the camera. As a LinkedIn reporter explained, "keep your virtual backgrounds professional and relative to your business" (Luff 2021). Yet, feedback often remained subjective. While explaining to LinkedIn audiences how to create a "rom-com office" with textile backdrops, "well-chosen books," and "twinkle lights," a media specialist explained that "Oh, and also FYI, virtual backgrounds are not appropriate for high-stakes Zoom meetings" (Kohn 2020). She emphasized that workers needed to create a "backdrop that doesn't look too deliberate or fussy. That allows you to stand out for what you're saying—not for how bad your setting is" (Kohn 2020). In other words, employees needed to mimic the walls of their workplace within their homes to be taken seriously in the Zoom borderlands. Commentators also emphasized controlling the sound coming from the microphone of employees' devices. As a commentator suggested, "[o]ne of the most important things you can do to maintain proper etiquette in Zoom meetings is to make sure you mute your mic when you are not speaking" (Weidinger 2020). In other words, employees needed to control visual and sound cues that reminded workers of the home, such as children, other working adults, and pets. Observers warned that non-work-related noises would cause professional "embarrassment" (Safronova 2020).

Many employers expected employees to set up a space in their homes for Zoom meetings that did not visually or auditorily reflect their personal lives. Columnists warned that if personnel did not control the background and sounds within their videoconferencing surroundings, they risked "distracting" other workers (Andersen 2021; Montgomery 2019; Weidinger 2020). As a Zoom commentator before the Covid-19 lockdowns explained: "By having a clean setting with work-appropriate art and decorations, you reduce the chance that attendees will get distracted" (Montgomery 2019). Another columnist clarified: "Seemingly small sounds and general background noises in your home bleeding into your business meeting may not seem like a big deal, but they can be a real distraction that is disruptive and disrespectful to whomever is speaking" (Weidinger 2020). These so-called intrusions had gendered and familial connotations, as they sought to eliminate from work the presence of home including chores, children, pets, and other adults, who might be seen or heard within the Zoom borderlands (Goldman 2020). For example, the presence of "laundry" within the digital box—whether in "piles," "dirty," or "unfolded"— caused concerns about workplace interruptions (Pennsylvania Society of Physician Assistants 2021; Dray 2020; Andersen 2021; Zight 2020). Historically, the washing of clothing has been associated with "shame" and relegated to the labors of women, especially women of color and working-class women (Fischer 2016, 821–43; Carson 2021, 11–23). By presenting visual signifiers of unkempt spaces such as laundry in the Zoom borderlands, workers revealed that family members, presumably women, left household labors undone. These chores, while essential for work, needed to be left "invisible" as if effortlessly undertaken (Ferrari 2020; Foster 2000, 9). Additionally, in July 2020, four months after the initial lockdown, a commentator for the *Wall Street Journal* encouraged employees to:

> Close the Office Door . . . Arrange for family to stay out of the way. At many places, pets and children are no longer the cute intrusions they were in the early days of the pandemic. Nobody wants to hear a housemate in the background unloading a dishwasher. (Morris 2020)

Another observer suggested that meeting distractions could come from sounds from outside the home made by the conditions of neighborhoods (Pennsylvania Society of Physician Assistants 2021). According to advice columnists, employees needed to provide the perception of control over space to maintain professional standards. In other words, employees needed to keep up the appearance that they could create separate spaces to compartmentalize work in their homes and simply "close the door" on what was happening elsewhere in their homes (Dray 2020). Yet, this prescription by employers ignored the social realities of gender, race, and class faced by employees in the home-work borderlands.

During the lockdown period, many workers' "personal lives [bled] into the workday" as they managed "taking time for childcare, virtual schooling, looking after pets" (Safronova 2020). Before the pandemic, many women worked the "second shift," which scholars explain as a gender phenomenon that occurred when working women also completed most of the housework and childcare needs at home (Brailey and Slatton 2019). During the pandemic, for many individuals, but especially working women, this imbalance of household labor expectations caused work and home tasks to be intermixed throughout waking hours (Ferrari 2020). As the public realms of work and school entered homes through the Zoom borderlands, colleagues often observed the second shift that some women faced along with other private matters in the home (Ferrari 2020; Wik 2021). When examining the second shift within an intersectional framework, unfinished housework and caretaking duties also pointed to the absence of domestic laborers that enabled many middle-class, and mostly white, women to pursue work and parenting while "outsourcing" house cleaning and childcare (Brailey and Slatton 2019, 31–3). During the pandemic, numerous domestic laborers—many women of color, who already faced labor exploitation around benefits and job security—lost essential income while also putting themselves and their families at risk. As reported, one domestic laborer "lost 50 percent of her clients" (Winkie 2021). Suggestions by business advocates for workplace professionalism in the home, during lockdown, prioritized the living conditions of those who had the space and in-home childcare arrangements to replicate office-like environments at home such as suburban homes as well as those from traditional nuclear families with a stay-at-home parent. Around this period, however, only "25% of children under age 15 living in married-couple homes had a stay-at-home mother, compared to only 1% with a stay-at-home father" (US Census Bureau 2020). Therefore, most married couple homes, as well as single-parent homes and split family households, did not meet these ideal workplace conditions and often struggled to meet these parameters or had to break with professional standards.

Commentator advice for controlling employees' behavior in private spaces also extended to concerns over employees' professional beauty standards while in the Zoom borderlands. Professional etiquette extended to the clothing worn by employees within and outside of the digital video box. Many observers took issue with employees wearing lounge pants termed "pajamas" (Andersen 2021). In a call to "make your appearance professional," a commentator explained that:

First things first, making yourself presentable for your videoconference is essential. You'd be surprised how many people show up to Zoom meetings in their pajamas or other attire that is completely inappropriate—from torn-up T-shirts and sweatpants to bathing suits while basking in the sun poolside, what people believe is acceptable for an online meeting can feel astonishing. The dress code might not be as strict as in the office, but you should still present a professional appearance. This is not limited to your clothing—you need to pay attention to your personal hygiene and grooming as well. (Weidinger 2020)

While some observers acknowledged that employers should have made allowances around clothing when the pandemic first hit, others explained, "as the world continues to go back to normal, you need to start paying close attention to what you wear during zoom calls" (James 2022). While both men and women often wore business tops and casual bottoms, calls for monitoring clothing worn in the Zoom borderlands remained gendered due to associations that soft casual clothing worn at home indicated "luxury," a lack of seriousness about work, and perhaps even stereotypes around femininity and stay-at-home mothers (Andersen 2021; Khazan 2020; Steiner 2007, x).

Many commentators, who called for controlling spaces and bodies to increase productivity and mitigate behaviors they deemed unacceptable, did not consider how these concerns around professionalism often led to racial and gender discrimination in the workforce. While many professionals have pointed to the racist and gendered implications of maintaining professional standards that already existed around clothing and hair, additional expectations around Black professional beauty standards remained in the Zoom borderlands (Buchanan 2020; Goldman 2020). As MSNBC reported in 2020, "Black women are 1.5 times more likely to be sent home from the workplace because of their hair" (Muller 2020). Black women also faced concerns about "be[ing] taken seriously" due to their hair choices (Muller 2020). As a professional Black woman explained:

> The debate around "presentability" as it pertains to Black hair is a frustrating and tiresome tug of war, with Eurocentric ideals on one end and expectations we've grown up with on the other. The concept that Black hair is wild and needs to be tamed is real, as evidenced by the fact that anti-hair discrimination had to be legislated, and still isn't outlawed in all 50 states (Clarke 2021).

Advice specifically for Black women recommended that "next time you have a lot of [Zoom] meetings planned for the day, consider trying out one of these dos: Updo, Updo with scarf, Ponytail puff or pineapple, Twists, Braids, Wash and Go," suggesting Black women still needed to put in additional effort to be considered presentable by their peers (*Instant Arẹ̀wà Hair* 2022). Yet some commentators indicated that in the Zoom borderlands, their hair preparations did not seem enough. As a professional explained:

> In a virtual space, with just my cute smile, unruly locs . . . I felt odd and spatially closer than I'd ever been to my colleagues. A little more exposed. A little more vulnerable. Zoom was magnifying everything, including my long locs, which I had formed into eight large braids to shape them into waves. My headband wasn't quite hiding their awkward thickness. (Buchanan 2020)

In addition to ways for wearing hair, correspondents suggested "avoid wearing distracting fabric patterns," pointing to specific cultural practices for some of the Black community such as bonnets and African textiles (Buchanan 2020; *Instant Arẹ̀wà Hair* 2022). As a commentator explained:

> At home or among our own, Black women, Latinas and other women of color will likely sport hoop earrings, hijabs and colorful patterns that feel natural for us. These garments, accessories, and symbols of culture and faith don't make us any less intelligent, confident, essential or professional. But they often mark us as different in institutions where implicit bias is a reality. (Buchanan 2020)

Other columnists described how this pressure to maintain professional expectations in the Zoom borderlands amplified discrimination against women of color, who often felt the need, due to systemic racism in the workforce, to wear certain clothing and hairstyles to have their work valued by their peers and managers (Buchanan 2020; Clarke 2021; *Instant Arẹ̀wà Hair* 2022).

Monitoring the professional beauty standards among workers, especially in a time of the pandemic, privileged those with the time, access to services, and emotional capacity to keep up standards deemed appropriate by business advocates. Women, especially women of color, often spent more time and money obtaining services and products to maintain professional beauty standards (Avila 2020). During the lockdown period, many working women lost access to salon services that they used to maintain professional standards. While some women reverted to box colors to "cover their grays," suggesting concerns over age discrimination among women, not all women had access to alternative solutions at home (Robin 2021). As a Black professional commenting on the discriminatory workplace expectations around hair explained:

> Add[ed] pressure to Black girls who become Black women with unsettled and uncomfortable relationships with their hair even before the complexities of the professional world set in. We walk through spaces hearing that we have to work

twice as hard to get half as much as our white colleagues, and that work includes styling our hair. Not everyone can just do wash-and-go's before an event—for some of us, wash day is an all-day routine that includes stretching, blow drying, parting, and twisting with a mixture of oils and creams dripping down our hands. (Clarke 2021)

As a pandemic solution and without access to hair services, some Black women wore hair bonnets in the Zoom borderlands, creating debates around professionalism. As one Black commentator explained:

> One thing I refuse to normalize is hoping [sic] on staff Zoom calls wearing my bonnet. . . . Bonnets are great for the house, but not the best for video calls with your colleagues and management. When a Zoom meeting is scheduled, and my hair isn't done, I cover my bonnet with a head wrap. This stylish alternative is a win because they can be worn both in the office and over a video call. (Badger 2021)

Despite having to find new workarounds, women across the Zoom borderlands expressed that their beauty practices mattered to other workers when presenting their professional persona.

Advice columnists often explained that this display of professional aesthetics should be done with the appearance of ease (Foster 2000, 9). One commentator explained that workers should "not use the camera as a mirror" (James 2022). On the surface, the writer situated this advice around "focus," but his comments indirectly implicated women concerned about professional beauty standards. He mislabeled these motivations as vain instead of recognizing workplace standards placed on women around beauty. As a scholar explained:

> Pressure to keep up appearances is largely directed at women. "So when men see themselves on a video platform, they don't look that different. The reality of the differential demands we put on women are just showing up in a different context." (Goldman 2020)

During this period, women faced higher levels of "Zoom fatigue" because of the workplace pressures around beauty standards tied in with seeing themselves as depicted on camera (Stillman 2021; Goldman 2020).

Many commentators also ascribed to the belief that employees' personal choices at home created a unique mindset that affected productivity within the Zoom borderlands. Columnists explained that wearing professional clothes from top to bottom established "the right mental space" for the job, which allowed for more productivity and connection when interacting within the Zoom borderlands (Andersen 2021). Employees needed to physically embody workplace culture in the home and on their bodies so that when they entered the Zoom borderlands they could labor effectively for their companies. Acting as a "go-between" in the Zoom borderlands, these employees used new tools such as stage lighting to illustrate compliance (Metcalf 2006, 1–16; Mull 2020). Other columnists viewed professional aesthetics such as clothing as essential to "hav[ing] better results" for the company and a signal to others about the employee's dedication level to their work (Fratangelo 2023; Andersen 2021; *Emily Post* 2020). For observers, the camera remained central for employees to present this dedication since, as "etiquette expert" Annette Y. Harris notes, "keeping the camera off signals that either you look like a mess or your plan is to multitask during the meeting instead of giving your full, undivided attention" (qtd. in Andersen 2021). Even though some observers made allowances for occasionally turning off cameras to avoid meeting disruptions, they warned that other modes of communicating to observers such as "body language," "gestures," and "eye contact" would be lost and could even contribute to "zoom fatigue." From their perspective, using tools in the Zoom borderlands increased workplace morale from the home, connecting professional aesthetics directly to driving results (James 2022).

New Customs in the Zoom Borderlands

Despite these expectations, women experienced a variety of challenges balancing work and home priorities in the Zoom borderlands. Despite columnists' suggestions of the opposite, many women experienced "Zoom fatigue" due to recurrent meetings on camera leading to "self-consciousness" (Stillman 2021). Scholars suggested that "those who are more worried they will be judged for their appearance, professionalism, or degree of engagement with the call spend more energy monitoring their own image during video calls, which leaves them feeling more drained afterward" (Stillman 2021; Goldman 2020). Other observers pushed back on these standards by explaining that "experts in body image and gender politics say that in this stress-rich, time-poor environment, the notion that women should be dolling up for virtual meetings . . . is simply the latest evidence of our society's tenacious, lopsided beauty standards" (Goldman 2020). Women of color explained the additional steps they had to take to reach professional standards that normalize the hair of their white counterparts. As one Black professional explained, "working from home should offer us the same sigh of relief as when we take off our bra or heels at the end of the day, but the fact that I still hustle to do my hair for my virtual meetings says the opposite" (Clarke 2021). In response, some women used videoconferencing software to resist these perceptions by creating sessions such as a "Zoom Bonnet Meeting"; as a show of support, women wore caps that safeguard their hair and discussed professional double standards for Black women (Clarke 2021). Other employees responded by purchasing special lighting, sometimes receiving approval to use company funds, to improve their look on Zoom calls "to be taken seriously" (Mull 2020). Individuals also felt pressure to invest in cosmetic surgery and dentistry as lockdowns lifted, citing the effects of viewing themselves via Zoom as a motivating factor (Advanced Cosmetic Dentistry 2020; Bell 2021). With the fear of being observed, professional beauty standards often operated at the lateral and self-monitoring level.

Other employees faced space and time constraints in their homes as they attempted to work around the schedules and needs of others. Many people found it difficult to find quiet places in the house that did not interfere with the work of others regardless of the lighting and background expectations. As one employee explained, after having an unexpected Zoom meeting while still in a bathrobe:

> And while the end result wasn't *actually* that big of a deal, it got me thinking about how we navigate this weird, new world where we're pivoting entire . . . business operations online. Where in-person meetings are now videoconference calls and the conference rooms are our living rooms or bedrooms or, occasionally bathtubs. (My girlfriend and I are living in a very small studio together and the bathroom is the only place with a door.) (Kircher 2020, 1)

For some employees who lived with more space per person, such as in the suburbs, they were able to devote specific areas for each member's work and school meetings. Yet those living in cities and with extended family in multigenerational homes felt these space constraints more acutely, citing the need for more space or flexibility around professional standards (Lautz 2021; Weik 2021). Limited space also became a concern for those workers, such as people of color, who operated different cultural identities at work and at home. As one commentator noted, "you get to see a slight snippet of your colleagues' houses which really helps you work out more about who they are and what they're about" (Dray 2020). This exposure, however, left many workers feeling "vulnerable" as they wanted to keep some aspects of their lives private from their coworkers, since many aspects of home life such as religious practices, cultural artifacts, sexuality, and personal hobbies could be observed and weaponized (Buchanan 2020; Dray 2020). As a Black professional explained:

> Zoom offered a window into our private spaces, and closing off mine, like some others had done, wasn't an option. Whether due to my older hardware or my lack of Zoom savvy, I was unable to substitute a still profile picture for my live image or to hide my environment with a cute virtual background. (Buchanan 2020)

Maintaining privacy and the perception of space neutrality while on Zoom often depended upon quickly learning how to use the application and having specific technical conditions sometimes made unreachable by class and digital literacy. Scholars have illustrated how similar "digital divides" can make low-income people, especially those who identify with intersecting social categories of race and gender, "hypervisible" while being surveilled (Eubanks 2011, 30, 36; Nash 2020). As a result, those in power create rules meant to punish those groups perceived as acting beyond professional social norms, while observers looked for behaviors in the Zoom borderlands that they deemed deviant based on their own discriminatory biases (Eubanks 2011, 30, 36; Nash 2020).

Some workers during the lockdown period, especially women, had to use the time that they usually devoted to work with caretaking duties. While children homeschooled using videoconferencing tools, they also needed more supervision from their parents, a task that often fell upon women (Heggeness and Fields 2020). Daycare and eldercare facilities also locked down during this period, adding pressure to those who identified as caretakers, again mostly women (Heggeness and Fields 2020; CDC Women's Health 2023). Many women faced the daily challenge of balancing caretaking needs while operating in their jobs (CDC Women's Health 2023; Eubanks 2011, 30, 36). Some women worried that employers and peers would negatively perceive their need to care for their children during working hours as seen in meetings as causing "distractions" and not contributing (Stillman 2021; Wik 2021). For many women, working in these extraordinary conditions meant lengthening their workday or not having a clear delineation between work and home, which increased exhaustion (Beheshti 2021; Wik 2021).

Many employees who identified as women often felt pressure to recognize the conditions of their home-work space. Often, they apologized to their coworkers for their appearance or the status of their home. Conversely, many employees justified why they turned off their cameras, and defended the rights of spouses, children, and pets to freely use their own home spaces, or make space for caregiving duties while on Zoom (Goldman 2020). Some workers began to resist calls for workplace oversight into the home by encouraging employees to show and tell their private lives in a public work setting to work toward gender equality in the Zoom borderlands. These well-intentioned calls, however, often entrenched other hierarchies, especially around cultural practices involving parenting, gender, and race in the process (Mason 2020; Wik 2021). Some etiquette writers suggested that if professional mishaps occurred then employees should communicate their situation for diverging from professional norms to the group since "people are being really forgiving about those sorts of things" (Kircher 2020, 2–3). Despite these reassurances of colleague congeniality, columnists clarified that work should be prioritized in the home (Kircher 2020, 1–6). Moreover, few in power actually questioned the workplace standards deemed professional as entrenching systemic discrimination, especially as they overreached into the homes of their employees that they now relied on. Instead, employees had to negotiate on a daily basis, based on their lived experiences, what rules they needed to follow in the Zoom borderlands.

This social policing in the Zoom borderlands that revolved around controlling space and bodies occurred within the context of a deadly pandemic. While business advocates attempted to control workplace standards in the home, employees, many of whom identified as women, faced extreme pressures due to professional standards, which were amplified by the gendered and racial systems in which they already worked. While some women complied with and even participated in the policing of professional standards while on Zoom, others used the opportunity to relax their own expectations either by choice or out of necessity. In other words, working women affected by the conditions in which they lived and worked made decisions based on their varying levels of agency on whether to follow or resist new norms manifesting in the Zoom borderlands.

References

Advanced Cosmetic Dentistry. 2020. "What Is the 'Zoom Boom'?" *Paul A. Peterson*, December 22, 2020. https://advancedcosmeticdentistry.com/blog/cosmetic-dentistry/what-is-the-zoom-boom/.

Andersen, Charlotte Hilton. 2021. "14 Zoom Etiquette Rules You Need to Follow." *Reader's Digest*, January 6, 2021. https://www.rd.com/article/zoom-etiquette/.

Avila, Allena. 2020. "How the Grooming Gap Could Affect You in the Professional World." *Her Campus*, January 25, 2020. https://www.hercampus.com/school/cal-poly /how-grooming-gap-could-affect-you-professional-world/.

Badger, Marsha. 2021. "To Bonnet or Not To Bonnet, That Is The Question: Are Hair Bonnets On Video Work Chats Considered Professional?" *HelloBeautiful*, January 13, 2021. https://hellobeautiful.com/3273610/hair-bonnets-on-work -video-calls/.

Beheshti, Naz. 2021. "We Worked Longer Hours During The Pandemic—Research Says We Need To Work Smarter, Not Harder." *Forbes*, August 18, 2021. https://www .forbes.com/sites/nazbeheshti/2021/08/18/we-worked-longer-hours-during-the -pandemic-research-says-we-need-to-work-smarter-not-harder/.

Bell, Maya. 2021. "A Zoom Boom for Cosmetic Surgery." *University of Miami Medicine*, Fall 2021. https://magazine.med.miami.edu/a-zoom-boom-for-cosmetic -surgery/.

Bender, Kelli. 2020. "Working from Home with a Pet? 4 Expert Tips to Keep Your Cat or Dog from Crashing Work Calls." *People*, October 23, 2020. https://people.com/ pets/expert-tips-to-keep-your-cat-or-dog-from-crashing-work-calls/.

Blum, Sam. 2022. "Employee Surveillance Is Exploding with Remote Work—And Could Be the New Norm." *HR Brew*, January 19, 2022. https://www.hr-brew.com /stories/2022/01/19/employee-surveillance-is-exploding-with-remote-work-and -could-be-the-new-norm.

Brailey, Carla, and Brittany C. Slatton. 2019. "Women, Work, and Inequality in the U.S.: Revising the Second Shift." *Journal of Sociology and Social Work* 7 (1). https://doi.org /10.15640/jssw.v7n1a4.

Brivot, Marion and Yves Gendron. 2011. "Beyond Panopticism: On the Ramifications of Surveillance in a Contemporary Professional Setting." *Accounting, Organizations and Society* 36 (3): 135–55. https://doi.org/10.1016/j.aos.2011.03.003.

Buchanan, Shonda. 2020. "Zooming While Black." *Sisters from AARP*, April 17, 2020. https://www.sistersletter.com/culture/zooming-while-black.

Carson, Jenny. 2021. "'We Win a Place in Industry': Black Women and the Birth of the Power Laundry Industry." In *A Matter of Moral Justice: Black Women Laundry Workers and the Fight for Justice*, 11–23. Champaign: University of Illinois Press. http://www.jstor.org/stable/10.5406/j.ctv1rnpjxf.5.

Clarke, Sydney. 2021. "It's Deeper Than the Bonnet: The Ever-Present Expectations Of Virtual Professionalism." *Refinery29*, January 7, 2021. https://www.refinery29.com/ en-us/2021/01/10217578/bonnet-natural-hair-work-zoom-appropriate.

Dray, Kayleigh. 2020. "19 Things You Only Know If You've Just Started Using Zoom." *Stylist*. https://www.stylist.co.uk/life/zoom-video-conference-calls-remote-meetings -working-from-home-webcam-fails-zoombombing/370736.

DuVal, Kathleen. 2010. "Borderlands." *Oxford Bibliographies*, May 10, 2010. https://www.oxfordbibliographies.com/display/document/obo-9780199730414/obo-9780199730414-0010.xml.

Emigh, Rebecca Jean and Corey O'Malley. 2020. "Face-to-Face with Zoom? Remote Teaching During the Coronavirus." *American Sociological Association*, August 14, 2020. https://asaculturesection.org/2020/08/14/face-to-face-with-zoom-remote-teaching-during-the-coronavirus/.

Eubanks, Virginia. 2011. *Digital Dead End: Fighting for Social Justice in the Information Age.* Cambridge, MA: MIT Press.

Ferrari, Aylen Rodríguez. 2020. "Why COVID-19 Pandemic Put Women's Second Shift in the Spotlight?" *Medium*, June 3, 2020. https://medium.com/@rguezf.aylen/why-covid-19-pandemic-put-womens-second-shift-in-the-spotlight-ab4be2466d24.

Fischer, Clara. 2016. "Gender, Nation, and the Politics of Shame: Magdalen Laundries and the Institutionalization of Feminine Transgression in Modern Ireland." *Signs: Journal of Women in Culture and Society* 41 (4): 821–43. https://doi.org/10.1086/685117.

Fornäs, Johan, ed. 2002. *Digital Borderlands: Cultural Studies of Identity and Interactivity on the Internet.* New York: Peter Lang.

Foster, Gwendolyn Audrey. 2000. *Troping the Body: Gender, Etiquette, and Performance.* Carbondale: Southern Illinois University Press.

Foucault, Michel. 1995. *Discipline and Punish: The Birth of the Prison.* 2nd Vintage Books ed. New York: Vintage Books.

Fratangelo, Nichole. 2023. "13 Golden Rules for Working from Home." *Reader's Digest*, January 19, 2023. https://www.rd.com/list/work-from-home/.

Goldman, Leslie. 2020. "The Inescapable Pressure of Being a Woman on Zoom." *Vox*, May 13, 2020. https://www.vox.com/the-highlight/2020/5/13/21248632/work-from-home-zoom-women-appearance-beauty-no-makeup.

Gray, Aysa. 2019. "The Bias of 'Professionalism' Standards." *Stanford Social Innovation Review*, June 4, 2019. https://ssir.org/articles/entry/the_bias_of_professionalism_standards.

Hämäläinen, Pekka and Samuel Truett. 2011. "On Borderlands." *The Journal of American History* 98 (2): 338–61. https://www.jstor.org/stable/41509959.

Haraway, Donna. 1988. "Situated Knowledges: The Science Question in Feminism and the Privilege of Partial Perspective." *Feminist Studies* 14 (3): 575. https://doi.org/10.2307/3178066.

Heggeness, Misty and Jason M. Fields. 2020. "Working Moms Bear Brunt of Home Schooling While Working During COVID-19." *U.S. Census Bureau*, August 18, 2020. https://www.census.gov/library/stories/2020/08/parents-juggle-work-and-child-care-during-pandemic.html.

The Henry Ford. n.d. "Ford Sociological Department & English School." Accessed May 2, 2023. https://www.thehenryford.org/collections-and-research/digital-resources/popular-topics/sociological-department.

James. 2022. "Zoom Etiquette: 15 Do's and Don'ts for Any Zoom Meeting." *Social Intents*, November 7, 2022. https://www.socialintents.com/blog/zoom-etiquette-15-zoom-meeting-rules-everyone-should-follow/.

Khazan, Olga. 2020. "Work from Home is Here to Stay." *The Atlantic*, May 4, 2020. https://www.theatlantic.com/health/archive/2020/05/work-from-home-pandemic/611098/.

Kircher, Madison Malone. 2020. "Zoom Etiquette from Emily Post's Very Chill Great-Great-Granddaughter." *Vulture*, March 27, 2020. https://www.vulture.com/2020/03/zoom-etiquette-best-coronavirus-video-chat-tips.html.

Kohn, Sally. 2020. "10 Beginner and Advanced Tips for Upping Your Zoom Game." *LinkedIn*, September 1, 2020. https://www.linkedin.com/pulse/10-beginner-advanced-tips-upping-your-zoom-game-sally-kohn.

Lautz, Jessica. 2021. "Full House: The Rise of Multi-Generational Homes During COVID-19." *National Association of Realtors*, February 23, 2021. https://www.nar.realtor/blogs/economists-outlook/full-house-the-rise-of-multi-generational-homes-during-covid-19.

Luff, Jim. 2021. "Zoom Etiquette - It's a Thing." *LinkedIn*, February 8, 2021. https://www.linkedin.com/pulse/zoom-etiquette-its-thing-jim-luff.

Maddox, Teena. 2020. "11 Ways to Be a Consummate Professional during Zoom and Microsoft Teams Meetings." *TechRepublic*, April 24, 2020. https://www.techrepublic.com/article/11-ways-to-be-a-consummate-professional-during-zoom-and-microsoft-teams-meetings/.

Mason, Jessica. 2020. "The New York Times Is Wrong, Everyone Wants to See Your Pet on Video Conferences." *The Mary Sue*, March 25, 2020. https://www.themarysue.com/the-new-york-times-is-wrong-about-pets-on-video-conferences/.

Metcalf, Alida C. 2006. *Go-Betweens and the Colonization of Brazil: 1500–1600*. Austin: University of Texas Press. https://doi.org/10.7560/709706.

Montgomery, John. 2019. "Video Meeting Etiquette: 7 Tips to Ensure a Great Attendee Experience." *Zoom Blog*, November 27, 2019. https://blog.zoom.us/video-meeting-etiquette-tips/.

Morris, Betsy. 2020. "Seven Rules of Zoom Meeting Etiquette from the Pros." *Wall Street Journal*, July 12, 2020, sec. Life. https://www.wsj.com/articles/seven-rules-of-zoom-meeting-etiquette-from-the-pros-11594551601.

Mull, Amanda. 2020. "Americans Got Tired of Looking Bad on Zoom." *The Atlantic*, November 19, 2020. https://www.theatlantic.com/technology/archive/2020/11/ring-lights-for-all/617143/.

Muller, Onicia. 2020. "What Does Looking 'professional' Mean Now?" *MSNBC.Com*, October 13, 2020. https://www.msnbc.com/know-your-value/what-does-looking-professional-mean-now-n1243196.

Muñiz, Ana. 2022. *Borderland Circuitry: Immigration Surveillance in the United States and Beyond*. Oakland: University of California Press.

Nash, Susan. 2020. "The Pandemic Has Accelerated the Need To Close The Digital Divide For Older Adults." *Stanford Center on Longevity*, June 4, 2020. https:// longevity.stanford.edu/the-pandemic-has-accelerated-the-need-to-close-the-digital -divide-for-older-adults/.

"The Only Guide to Zoom Call Etiquette You'll Ever Need." 2020. *Zight*, September 3, 2020. https://zight.com/blog/zoom-call-etiquette/.

Perales, Monica. 2013. "On Borderlands/La Frontera: Gloria Anzaldúa and Twenty-Five Years of Research on Gender in the Borderlands." *Journal of Women's History* 25 (4): 163–73. https://doi.org/10.1353/jowh.2013.0047.

"Professional Black Women's Secrets to Looking Zoom-Ready." 2022. *Instant Arẹ̀wà Hair*, June 28, 2022. https://www.instantarewahair.com/blogs/news/professional -black-women-s-secrets-to-looking-zoom-ready.

Robin, Marci. 2021. "Here's How the Pandemic Changed Our Hair Colors for Good." *Allure*, January 15, 2021. https://www.allure.com/story/pandemic-hair-color-at -home-kits.

Safronova, Valeriya. 2020. "Digital Hygiene in the Zoom Era." *The New York Times*, October 22, 2020, sec. Style. https://www.nytimes.com/2020/10/22/style/zoom -safety-protocols.html.

Steele, Catherine Knight. 2021. *Digital Black Feminism. Critical Cultural Communication*. New York: New York University Press.

Steiner, Leslie Morgan. 2007. *Mommy Wars: Stay-at-Home and Career Moms Face off on Their Choices, Their Lives, Their Families*. New York: Random House Trade Paperbacks.

Stillman, Jessica. 2021. "Why Women and New Employees Are More at Risk for Zoom Fatigue, According to New Research." *INC*, October 28, 2021. https://www.inc.com/ jessica-stillman/zoom-fatigue-burnout-women-new-employees.html.

Taurek, Bernhard H. F. 2019. "Surveillance: A Complex Relationship." In *Surveillance - Society - Culture*, edited by Florian Zappe and Andrew Gross, 27–38. New York: Peter Lang.

US Census Bureau. 2020. "Census Bureau Releases New Estimates on America Families and Living Arrangements." https://www.census.gov/newsroom/press-releases/2020/ estimates-families-living-arrangements.html.

Weidinger, Steven. 2020. "What Is Proper Zoom Etiquette?" *Washington Post Jobs*, September 28, 2020. https://jobs.washingtonpost.com/article/what-is-proper-zoom -etiquette-/.

Weik, Taylor. 2021. "Multigenerational Homes Are a Source of Stress, Comfort During COVID-19." *Teen Vogue*, February 17, 2021. https://www.teenvogue.com/story/ multigenerational-homes-covid-19-pandemic.

Wik, Tracey. 2021. "3 Ways to Rethink Outdated Workplace Practices That Stigmatize Working Mothers." *Fast Company*, October 12, 2021. https://www.fastcompany .com/90685108/3-ways-to-rethink-outdated-workplace-practices-that-stigmatize -working-mothers.

Winkie, Luke. 2021. "What Domestic Work Looks and Feels Like Right Now." *Vox*, January 15, 2021. https://www.vox.com/the-goods/2021/1/15/22225692/national -domestic-workers-alliance-us-covid-19-rights.

"Women, Caregiving, and COVID-19." 2023. *CDC Women's Health*, March 7, 2023. https://www.cdc.gov/women/caregivers-covid-19/index.html.

"Zoom Etiquette: Tips for Better Video Conferences." 2020. *Emily Post*, December 12, 2020. https://emilypost.com/advice/zoom-etiquette-tips-for-better-video -conferences.

"Zoom Meeting Etiquette: 15 Tips and Best Practices for Online Video Conference Meetings." 2021. *Pennsylvania Society of Physician Assistants*. https://pspa.net/wp -content/uploads/2021/03/Zoom-meeting-etiquette.pdf.

Staging Zoom

The chapters in this section bring to the fore questions of performance and the frame in which these acts occur, exploring Zoom not only as medium and interface but also as a *place* of and for performative acts. We use the term *staging* here to call attention to the specific spatial arrangements that Zoom establishes and the ways in which social actors must orient their performances in a manner dictated by the affordances and constraints of that space.

In gesturing toward affordances, we are taking seriously the degree to which media operate as environments and, in doing so, establish parameters for potential action. Such an acknowledgment does not, in itself, suggest a technological determinist view of human-machine interactions; rather, it acknowledges how a medium operates as a structure of potentials within which individuals likewise explore and exploit potential modes of performance. Such a perspective has more in common with Goffman's (2008) discussion of adapting social performance within shifting social contexts than it does McLuhan's (1994) analysis of media as sensory extensions (and amputations) of "man." If we take seriously the *place* of Zoom as a social environment, we can begin to appreciate the complexity of the performative context that it offers. On one hand, the logic of Zoom naturalizes the corporate-style meeting for all sorts of personal and interpersonal interactions, from Zoom births (Kaufman 2021) to deathbed vigils (Sci, this volume). At the same time, however, videoconferencing platforms such as Zoom *denature* domestic spaces as they transform everything that falls within the frame of the medium into an environment for institutional performance. When the world shut down during the Covid-19 pandemic, and all forms of social interactions shifted to digital spaces, the performative demands of working from home became all the more complex in this merging of private and public spaces. The hyperdependence on Zoom for all modes of interaction meant we all experienced—and perpetrated—forced entry into spaces often kept private. In such an environment, resistance can take on myriad forms, including a refusal to take up those potentials afforded by the digital interface—a topic we address in greater detail in our contribution to the final section of this collection (see also Cirucci 2023).

We begin with John A. McArthur's chapter, "Proxemics and Nonverbal Communication Dilemmas on Zoom," which explores how Zoom mediates nonverbal communication through the lens of proxemics, the study of the use of embodied space. Here, McArthur unpacks how we perform, and are perceived, through Zoom via unspoken presentations of self and digital interactions. By examining nonverbal interactions and connections via proxemics, McArthur urges us to consider how we might change the ways in which videoconferencing tools are used for teaching, learning, and social connections. Understanding one's relation to embodied space, and how the frame of Zoom mediates an interpersonal space, sets the stage (if you will pardon the pun) for the remaining two chapters in this section.

From a discussion of proxemics, we turn to a consideration of the affordances of videoconferencing with an eye toward both the platform and the medium as an aesthetic opening. The "pivot" that McArthur describes within the performative environment of the classroom likewise occurred on other more literal stages, as artists and performers saw a massive loss of space, audiences, and income throughout 2020. The range of bands and musical artists who engaged Zoom and found a potential for a mediated arts practice extended from small, indie bands to arena-packing pop stars (Billboard 2021). While many television shows were forced into production delays, live television—most notably news and talk shows—found in Zoom and other modes of videoconferencing technology a means of production that not only allowed for social distancing but also capitalized on—in every sense of the word—its affordances (Wagmeister 2020). In a particularly successful negotiation of these complex frames, for example, Stephen Colbert famously addressed his *Late Show* audience, less than a week after the WHO's pandemic declaration, "dressed in his usual suit while soaking in the bubble-filled tub" (Miller 2020). Likewise, the BBC found a means to embed the affordances of videoconferencing and the simultaneous restrictions on live theatre performance into a successful "videocall series" entitled *Staged*, starring David Tennant and Martin Sheen (Molano 2022). Daniel O'Brien's contribution to this section, "Zoom's Performative Window: Affordances and Constraints," enters into this conversation through an exploration of the remediation of film and theatre through four artistic pieces that were created and performed over Zoom: the short films *Magnolia Zoomed* and *Slap that Bass Zoomed* by Ian Garwood, Laurence Owen's autobiographical solo show *Some Other Mirror*, and Nic Blower's documentary *WE'RE IN THIS TOO*. O'Brien uses these pieces to consider how Zoom can create both closeness and intimacy through some performances and experiences, while still creating distance that can be enforced, as he notes, through "a click of a button."

Finally, Craig Fahner's chapter, "*Eigengrau*: Reimagining Videoconferencing as a 'Slow Platform,'" explores how artistic engagement with a medium can reveal both the embedded logic of the platform and the cracks within that logic: exploits (Galloway and Thacker 2007) that suggest otherwise unthinkable modes of use and means of engagement. With this in mind, Fahner developed *Eigengrau*, a videoconferencing platform that challenged the *video* itself and its epistemological primacy. What would "engagement" consist of if a platform required its users to *close their eyes*, and how might the sensorium of sound upend the implicit and explicit gaze of the camera that frames the Zoom frame itself? Such experiments with aural space as a medium for connection align well with Della Ratta's (2021) observation that "the disembodied sound dimension, and its potential for a freeing-up from the anxiety of the hyperconscious self-gaze, might help create an opening for spaces of intimacy and in defiance of connectivity. . . . Connectivity does not always equal connection—and vice versa." While small-scale, experimental platforms such as *Eigengrau* have little chance of disrupting the market dominance of corporate players like Zoom, these interventions, at the level of aesthetics as well as code, suggest an opening beyond the closures of a performative logic and that another platform – and more ways of performing, engaging, and connecting – is possible.

In each of the chapters in this section, the authors offer an understanding of the logic of Zoom in a performative context, one that asks us to keep in mind the relationship between actor and environment, whether that staging is a social performance or an artistic act. This emphasis on the co-relation between actor and environment, and the affordances of the medium, creates a space to consider how not all choices are foregone conclusions, even within network structures that support a dominant logic of a network society.

References

Billboard. 2021. "Here Are All the Livestreams & Virtual Concerts to Watch During Coronavirus Crisis (Updating)." *Billboard*, January 26, 2021. https://www.billboard.com/music/pop/coronavirus-quarantine-music-events-online-streams-9335531/.

Cirucci, Angela M. 2023. "Zoom Affordances and Identity: A Case Study." *Social Media + Society* 9, no. 1. https://doi.org/10.1177/20563051221146176.

Della Ratta, Donatella. 2021. "Teaching Into the Void: Reflections on 'Blended' Learning and other Digital Amenities." *INC Longform*, January 6, 2021. https://networkcultures.org/longform/2021/01/06/teaching-into-the-void/.

Galloway, Alexander R. and Eugene Thacker. 2007. *The Exploit: A Theory of Networks*. Minneapolis: University of Minnesota Press.

Goffman, Erving. *The Presentation of Self in Everyday Life*. New York: Anchor Books, 2008.

Kaufman, Gil. 2021. "Kelly Rowland Reveals Beyonce & Michelle Williams Watched Her Give Birth on Zoom." *Billboard*, April 30, 2021. https://www.billboard.com/music/pop/beyonce-watched-kelly-rowland-give-birth-zoom-9565694/.

McLuhan, Marshall. 1994. *Understanding Media: The Extensions of Man*. Cambridge, MA: MIT Press.

Miller, Mike. 2020. "Stephen Colbert Delivers Surprise *Late Show* Monologue from Bathtub." *Entertainment Weekly*, March 17, 2020. https://ew.com/tv/stephen-colbert-surprise-bathtub-monologue-late-show/.

Molano, Jessica Camargo. 2022. "Theatre as a Means of 'Interpreting' Lockdown: The Case of *Staged*." In *Theater(s) and Public Sphere in a Global and Digital Society, Volume 1*, edited by Ilaria Riccioni, 177–85. Leiden: Brill. https://doi.org/10.1163/9789004529816_015.

Wagmeister, Elizabeth. 2020. "The (Talk) Show Must Go on: How Daytime TV Has Safely Continued Production Amid the Pandemic." *Variety*, November 18, 2020. https://variety.com/2020/tv/news/talk-shows-covid-19-drew-barrymore-wendy-williams-1234834263.

4

Proxemics and Nonverbal Communication Dilemmas on Zoom

John A. McArthur, Furman University

At the onset of the Covid-19 pandemic, I was part of a team at our university that helped to lead the strategic redesign of learning for our classrooms. Over that fateful spring break, the "classroom pivot," as it came to be called, resulted in the mass migration of all coursework from in-person settings to synchronous Zoom-based classrooms (Haywood et al. 2023). As a scholar and researcher of physical and digital gathering spaces, I couldn't resist the urge to study this shift from the lens of proxemics.

Proxemics, the study of human use of space, is a term coined by anthropologist Edward T. Hall in his seminal work *The Hidden Dimension* (1966) and applied in many later works. Hall observed happenings like pigs feeding at a trough and birds sitting on a wire and applied those same principles to humans organizing themselves in queues or spacing themselves around a table. The hidden dimension he referenced is that unseen yet fully known pattern of how we place ourselves in relation to other people. Deep observation and study revealed that people use space differently based upon factors like gender, height, cultural patterns, interpersonal liking, body orientation, and relative body heat, among many others. Additionally, Hall chronicled the use of sensation and perception as indicators of interpersonal distance. He articulated the means whereby perceived visual detail, hearing, haptics (the ability to touch), thermal reception, and olfactics offered humans the information needed to determine how to negotiate the use of space with copresent others.

Hall's (1966) study and its later applications offer a few key ideas central to the investigation of Zoom from the perspective of proxemics. First, proxemics is grounded in the categorization of four distance zones: intimate, personal, social, and public (Hall 1966, 1968). Intimate distance references the closest

18 inches to the body. Two copresent individuals negotiating an encounter at intimate distance are often but not necessarily touching each other. They might be sharing a welcoming hug, whispering to each other, beginning a romance, or outright fighting one another. Hall titled the space between 18 inches and 3 feet from the body as personal distance. Individuals encountering each other at personal distance trigger a response. If the copresent other is a friend, we might welcome them into our personal space with a nod, smile, eye contact, or a verbal greeting. Were the other a foe, our response might be quite different. The entrant may be greeted with a warning, a blocking motion, a step backward and away, or another fight or flight response. Beyond personal distance, at the distance 3—6 feet from the body, is social distance. We engage most of our interactions at social distance. This distance is comfortable for conversation, greeting, meetings, or other social interactions such as exercising, working, or dining together. Beyond social distance exists public distance. Individuals come and go in public distance with varying degrees of interest. From 6 to 12 feet, which Hall titled close public distance, we survey entrants and move toward or away as needed. From 12 to 25 feet, which Hall titled far public distance, we recognize known persons and evaluate unknown persons. Beyond 25 feet, we tend to perceive less information than we need to engage in an interpersonal interaction with another person. Hall noted that these distances are culturally defined and that even though all humans engage in these categories of interpersonal distancing, the relative boundaries might shift marginally across cultures.

Second, Hall (1966) noted the role of sociospatial design in the negotiation of space. Sociospatial design refers to the orientation of people in space as a predictor of interaction possibilities. Consider a drawing room with a fireplace and seating for eight people. A typical arrangement of seating in this room might be to place all furniture in a semicircular or rectangular formation around a fireplace. This design suggests that all eight people will face the center and thereby encourages all eight to interact with each other. This arrangement is called a sociopetal arrangement because it orients people in the room toward each other. An unlikely arrangement of furniture in the room would be to place the furniture so that each person would face a section of wall with their backs to center. This design, called a sociofugal arrangement, would discourage interaction between copresent others. While this might not be a preferred seating solution in a drawing room, it might be a very apt choice for a library, work room, or study space.

Sociospatial design is not as binary a concept as my earlier description suggests. Most often designers of spaces employ a mixture of sociopetal and sociofugal arrangements to achieve a desired interaction outcome. In our drawing room, two chairs might be placed near the fireplace, with six others around a table or ottoman to create varied interaction possibilities, allowing for quiet conversations and group interactions simultaneously. A better example of this occurs in a restaurant where tables are oriented so that guests who arrive together are seated in a sociopetal arrangement to one another but in a sociofugal arrangement from other parties. This combination of sociopetal and sociofugal arrangements creates social space for each dinner party in a larger public space designed to accommodate multiple dinner parties at once.

In his book, *Personal Space: The Behavioral Basis of Design*, Robert Sommer (1969) expanded these notions of sociospatial design to incorporate other types of space and interactions, including classrooms. He famously noted that "teachers are hindered by their insensitivity to and fatalistic acceptance of the classroom environment" (Sommer 1969, 119). This notion was realized all the more in the Covid-era classroom as teachers lamented the sudden shift in learning environment and were forced to try a new approach, fraught with preconceived notions of the classroom and how learning occurs best. Interestingly, this idea was repeated by several instructors responding to the survey I conducted in 2020 (cited later; see also McArthur 2021 for a full discussion), including one respondent, who wrote:

> Whether zoom works well or not depends on the approach taken by the professor. If we try to make it a direct substitute for what we do in person, then we are missing the point and missing an opportunity to think more creatively about pedagogy. Zoom is fine during a crisis and we would all do well to adapt to it with some enthusiasm instead of complaining about it as a reality. Things will not always be this way and I think that this interruption of the norms of teaching might end up being beneficial for breaking the pedagogical complacency of far too many professors. (McArthur 2021, unpublished data)[1]

This respondent captured the mood of many peers whose optimism overtook their fear. However, this respondent's answer also belies the broad perception that Zoom was less ideal than a physical classroom.

Third, in my own reimagining of Hall's concepts for digital environments, *Digital Proxemics: How Technology Shapes the Ways We Move*, I furthered the study of proxemics and this notion of sociospatial design for a digital age through the exploration of "proxemospatial design" (McArthur 2016). Proxemospatial design suggests that digital interactions add an additional layer of information to sociospatial design. This layer of information either connects us to a location (proxemopetal design) or pulls us away from a location (proxemofugal design). An augmented reality interface in which a layer of light and sound is added to a building in a performative setting might create a proxemopetal design, connecting visitors more deeply to that physical location. Conversely, a virtual reality headset that creates an immersive digital experience creates a proxemofugal interaction with a user's immediate physical setting. Like sociospatial design, proxemospatial design might be best employed when proxemopetal and proxemofugal arrangements are mixed together to create meaningful interpersonal interactions.

This mixing of proxemospatial arrangements is grounded in some of the early work on interaction design in user experience design studies, most notably *Windows and Mirrors* by Jay David Bolter and Diane Gromala (2003). In their book, Bolter and Gromala suggest that the computer functions as both a window to the information presented on the screen and as a mirror to the capabilities of the program interface on the device. They point out that the window is most realized when users are immersed in the information and forget that they are peering into a computer. Conversely, the mirror is most realized when the interface causes a dysfunction in access through a glitch, latency, or lack of intuitive interface design features. The mirror also functions to reflect the user, noting differences between users as the device responds to the commands of the user (Bolter and Gromala 2003). This situation, which was figurative for Bolter and Gromala, is quite literal on the Zoom platform, since the user is at once peering into the information on screen and fully aware of the platform's presence. Thus, technologies in the genre of Zoom function as both sensing and responsive technologies. They sense and input human intent and relay visual and auditory stimuli to digitally present others. These actions cause us to see the technology and to see through the technology.

These ideas grounded in proxemics—interpersonal distance, sociospatial design, proxemospatial design, and interaction design—connect and become intertwined in the study of emerging digital, hybrid, virtual, and augmented gathering spaces. In the Zoom interface, the platform itself functions as both a physical space of encounter in the user's physical environment and a digital space of encounter in the compiled design of the Zoom gallery. Even though Zoom is one of these spaces among many, the ubiquitous use of Zoom during the pandemic and its rapid adoption created an ideal moment to study users' nascent responses to proxemics and nonverbal communication tactics in this digital gathering site.

Four Nonverbal Dilemmas

In January of 2021, my article on the four nonverbal dilemmas faced by instructors on Zoom was published in *Communication Teacher* (McArthur 2021). Even though the methods and outcomes of this analysis were presented fully therein, I will offer a succinct overview of the methods here. I surveyed 351 instructors who made the pivot from a seated classroom to a Zoom classroom across educational sectors. Participants reported experiences at every grade level from kindergarten to doctoral work and in a diverse range of subjects and disciplines. The survey invited respondents to reflect on their nonverbal interactions in the Zoom space as they related to proxemics, including the movement of their bodies, their facial expressions and eye contact, and their use of space and the artifacts within it. The resulting data contained 860 written statements about a wide variety of nonverbal experiences on Zoom. I employed grounded theory methodology to explore the data, allowing themes to emerge from the data analysis (Glaser and Strauss 1967). The data suggest that these instructors faced noteworthy dilemmas that warranted negotiation in each classroom (Zoom room) setting. The dilemmas they described are fairly generalizable to all Zoom users as we seek to navigate our communication identities in its synchronous, digital space. This chapter seeks to expand and deepen the concepts presented in McArthur (2021), offering the opportunity to explore and directly examine each dilemma more deeply in the context of proxemics.

The four nonverbal dilemmas that emerged within that study are *animating, replicating, reciprocating,* and *self-monitoring*. *Animating* refers to the performative act of presentation and the dilemma of a felt need to exaggerate nonverbal behaviors in the Zoom space. *Replicating* refers to the dilemma facing instructors as they attempt to maintain the same nonverbal behaviors in the Zoom space as in a seated classroom. *Reciprocating* refers to a dilemma surrounding how to give and receive nonverbal feedback during interactions on Zoom. *Self-monitoring* suggests the dilemma inherent in watching oneself while speaking in the Zoom interface. Note that each of these four dilemmas is presented in the present participle form of the verb. This distinction is purposeful because these are active dilemmas that impact the user in real time. Participants in Zoom environments continuously negotiate these dilemmas with varying degrees of success. And, they use different strategies to navigate each dilemma based upon the situation facing them in a given moment. These actions are labeled dilemmas to indicate that they are solvable negotiations without fixed outcomes. As Hall (1974) pointed out, interpersonal interactions follow generally recognizable patterns, but each specific interaction contains unique individuals with unique characteristics that can change the method of negotiation in that particular dyad or group. The same is likely true in the Zoom environment.

Additionally, each of these four dilemmas was present in the data across the nonverbal forms of communication used to prompt participants in the study. When asked about the specific visual elements of nonverbal communication— kinesics (body language including gesture), facial expression and eye contact, artifacts, proxemics, and physical appearance—in separate prompts, the same four themes emerged from the instructor responses about each of the nonverbal forms. This suggests that the negotiation of nonverbal communication is similar across the variety of nonverbal, visual cues employed by participants in the Zoom environment. In the descriptions to follow, note that the participants' word choice and grammar are repeated here to capture their original thoughts precisely and to relay their writing as presented in the data.

Animating

Animating was the first and most robust dilemma described by participants. In the earlier study, animating referred to "an instructor's purposeful choices to use or exaggerate nonverbal behaviors in the Zoom classroom to illustrate emotion, identity, or course content" (McArthur 2021, 5). One respondent described their actions like this: "I exaggerate my smiles when I praise them, and try to make sympathetic faces when they struggle. When I give demonstrations, I make 'think' faces, and 'eeek' faces; I have been filmed and it is a little bit embarrassing how much my face is exaggerated on Zoom." Another wrote, "I'm not sure it is a strategy, but I find myself exaggerating my facial expressions and responses to comments. Big smiles, nodding, bigger gestures, etc. All just ways of signaling I'm listening and enthusiastic." These instructors relied on facial expression to highlight and emphasize key points and ideas in the class content and interpersonal relationship. Notably, many comments referenced the purposeful exaggeration of these facial expressions for the camera in the Zoom platform.

This category of comments contained many other nonverbal forms. As you might imagine, some of my favorites involved the use of the camera to purposefully alter perceptions of interpersonal distance through animation. One respondent wrote, "I move from side to side, close and away, to make it more dynamic. It is doable even though the zoom window is small. One time, I whispered to my headphone, saying 'Can I tell you a secret?' More than half of my students leaned toward their computer screens, like cats." This image of the students leaning into the computer is a key indicator that the instructor successfully created a moment in which the Zoom interface functioned as a window to the digital gathering site and immersed the participants in the content at hand. Even though we cannot know whether they were leaning toward their device's speaker to hear the comment or they were moving closer to the instructor interpersonally, the visual and proxemic effect was the same—a shift from personal distance toward an intimate distance in which secrets can be shared.

Another respondent wrote about proxemics, humorously noting, "[I] get very close to the camera to call their *atención*. They can see only my open eye!" This instructor's dramatic transition from the display of personal distance to the display of intimate distance should have created some response from the others in the digital space. This boundary violation in physical space might have resulted in a verbal response, laughter, or physical movement, such as backing away from the invader. On Zoom, one would expect to see varied reactions, depending on the type of gallery or user window being employed by each student on their own devices.

This comment also reveals an additional concept about the role of both camera and gallery in the Zoom platform: cameras position us all at personal distance rather than social distance (the distance we are most accustomed to in a classroom setting). This acknowledgment is a fascinating bit of data. It speaks to the question of comfortability in the Zoom platform and suggests that for some people, the vantage point of personal distance designed into the Zoom platform is a threatening distance for interactions that would typically occur at social distance. This idea furthers the dilemma signified in animating, and various respondents negotiated the dilemma quite differently from their peers. One instructor wrote, "I have my screen and external webcam set up so that I don't appear too close to them. Cameras can make us feel too close to people and that's stressful. I set mine back a little to give people some personal space." The logic of the Zoom platform places us squarely within personal distance with our digitally copresent others, which may not always be the interpersonal distance of best fit.

Still others commented on the realization of their confinement to the camera space. One respondent offered, "my gestures and movements are bigger for those on Zoom, although limited to whatever space is on camera." This respondent reminds us of the affordances and constraints present in the Zoom interface and its function as both window and mirror. Their typical gestures were constrained by the visual frame of their selected camera. The hardware functioned as a forceful limitation on nonverbal behavior. Interestingly, the various respondents addressed this limitation in different ways by employing active motion of the body toward and away from the camera.

Additionally, animation was also referenced through the use of physical appearance and artifacts shown on camera. Comments on the use of eyebrows, facial hair, and makeup (of both the physical and digitally augmented varieties) suggested that respondents purposefully manipulated their appearance with both physical techniques and using Zoom's facial augmentation tools (both of these augmentations were released as part of Zoom's Studio Effects rollout in September 2020, in the middle of the course/semester in which the respondents for this study were teaching). Artifacts referenced in the data included visual aids, physical props, clothing, hair styles, digital flowers appended above the ears, as well as unique teaching aids like cats and potted ferns. For example, one respondent commented, "I have a fern in a face pot. From time to time, I use Dr. Fern for questions and answers. It's silly. But it allows me to answer what I know are FAQs in this level class. It gets a reaction from students, sort of like Dad jokes. Even the people who roll their eyes at Dad jokes still laugh at them." I can imagine this instructor carrying on conversations with Dr. Fern in each class, voicing each side of the conversation and enlivening the discussion.

These uses of facial expression, gesture, proxemics, physical appearance, and artifacts are each indicative of the role of nonverbal communication as an animating agent in the Zoom environment. Zoom complicated the performative nature of teaching and required the instructors to develop new methods for engaging students through nonverbal communication. Even so, each instructor addressed this dilemma differently, opting to employ the nonverbal technique to an exaggerated degree, to use it as they might normally, or even to diminish its use to create a less threatening learning environment. Whether these uses were intentional or only noteworthy upon reflection is a conversation we will return to shortly.

Replicating

The dilemma of replicating was referenced in responses throughout the study and defined as "an instructor's attempt to use nonverbal communication to maintain a sense of stasis between their own teaching experience in a physical classroom and their teaching experience in a Zoom classroom" (McArthur 2021, 6). This sense of replication was both an indicator of the necessity of the shift to Zoom and the work each of us performs in relation to our own identity maintenance. We strive to be the same person in dissimilar environments. Interestingly, this dilemma was represented both by those trying to keep everything the same regardless of their physical environment and those who shifted their approaches in dramatic ways or adopted new strategies to maintain similar levels of classroom engagement.

When commenting on the use of gestures, one respondent wrote, "I do this anytime I am teaching—Zoom doesn't change anything. We are bodies, we teach and learn with our bodies, not just with words." This idea is an interesting and perhaps typical response to the replicating dilemma, affording the individual the space to maintain their own identity regardless of the space of interaction. But it also neglects the material differences between a digital Zoom platform and a seated classroom. The Zoom environment is a physically different space with different environmental factors than the space of a seated classroom environment (which themselves can vary widely in form). Instructors' adaptability to classroom spaces—or "fatalistic acceptance" of classroom spaces (Sommer 1969, 119)—is perhaps a typical teaching dilemma that is addressed through strategies of replication. Still, I would speculate that the adaptation of functional, nonverbal communication from one environment to another could create tension for individuals engaged in communication practices in new spaces.

Like the first respondent quoted earlier, many others described strategies of replication as purposeful, nonverbal interventions in their work on Zoom. About gestures and body movement, one instructor noted, "I communicate on Zoom a great deal the same way as I communicate in person. I just try to keep my gestures inside of my zoom box." Another offered, "I don't do this any more or less often than I would in the classroom. I move around a lot and use gestures to demonstrate concepts often." Even while noting the difference in communication environment, these respondents remarked on their successful ability to replicate their nonverbal strategies across communication environments.

Such commentary was not only limited to kinesics, gestures, and body language but also referenced other nonverbal forms. One respondent wrote, "I always wear makeup and a shirt that looks camera friendly/professional. I think of replicating the classroom experience." About the artifacts used in the classroom, another respondent noted that their strategy was "dressing professionally—at least those parts that are visible on the screen—to remind students that we remain engaged in a serious pursuit, even though we're all in our homes or dorm rooms." These rhetorical claims suggest that the Zoom classroom was somehow lacking in the ability to create a professional or serious environment, as the instructors noted the need to remind their students of this aim for the classroom. Furthermore, this response indicates the participant's desire to maintain stasis, the central characteristic of the replicating dilemma, even during a pandemic.

Rhetorical claims like this appeared to undergird the responses surrounding replication. In some cases, these claims illustrated a denial of differences in the synchronous seated classroom versus the synchronous Zoom classroom. Other times, these claims emerged as articulations of confidence that I, the professor, was capable of providing a valuable learning experience on Zoom just as I would be in my seated classroom space. Still other times, these claims suggested a sense of collective perseverance that if we could move through the Covid-19/Zoom era, classroom environments could return to normal. An interesting fourth rhetorical claim also emerged suggesting that the past experiences of instructors in online teaching served as a predictor of success in the Zoom classroom.

In each of these four cases, the rhetorical claim articulated by the individual led them to reinscribe their work as instructors onto the digital Zoom environment, creating a scenario ripe for replication of behaviors. These types of rhetorical claims may have been unusually strong, given the context of a pandemic-induced, forced pivot to a Zoom environment. But they may also be robust articulations of our identities as teachers and learners, engrained within particular pedagogies for our subject matter. Likewise, this concept of replicating may take root across industries and situations, inspiring innovation and change in some people and entrenching others in tried-and-true methodologies—hence the dilemma. The replicating dilemma demonstrates this tension between stasis and adaptation, which seems to be characterized as adapting the classroom to Zoom delivery alongside the rhetorical claim that reminds us that "nothing has changed." Is it this stated reminder that creates the opportunity for replicating?

Scholars studying online social interaction during the pandemic have also taken on this dilemma of replication to different ends. In *Christian Century*,

Pastor Chris Palmer (2022) authored a short piece titled, "Bodies in Silence: A Worship Practice Zoom Can't Replicate." In the piece, Palmer argues that people need to be physically copresent to engage in meditative corporate worship, stating, "these are habits that only proximity will make possible—habits often discouraged by the disciplines of our digital age," suggesting that Zoom fails to capture the nonverbal communication form of silence (Palmer 2022, 13). Conversely, doctoral student Joanne Marras Tate (2023) studied the ability of synchronous videoconferencing to replicate "humannature" connections in aquariums left vacant during the pandemic. In *Environmental Communication*, she reported on her study of social isolation. Connecting viewers to the garden eel exhibit via video appeared to replicate the experience of human exposure for garden eels who had become reliant on such interaction during their life in the aquarium. In this case, the Zoom platform could recreate some forms of nonverbal communication that had been lost. The diversity of situations and the response to the ability of Zoom to replicate communication remains a dilemma that encourages users to consider multiple strategies for diverging ends.

Noting that they would attempt to replicate the outcome of their craft rather than the method, this respondent can have the final word on replicating: "I don't think of [nonverbal communication] as strategy so much as human connection. I'm a pretty animated person. Being on a computer doesn't change that."

Reciprocating

Reciprocating refers to "an instructor's desire to use nonverbal behaviors to give and/or receive feedback from students in the Zoom classroom, and the dilemma faced by instructors about how to create transactional communication in the Zoom space" (McArthur 2021, 7). This concept of reciprocating illustrates our recognition that communication involves complex patterns of turn-taking, and that turn-taking is largely dependent on nonverbal cues. Whereas responses concerning the two previous dilemmas offered a variety of measured responses, participants were clearly polarized on the response to the dilemma of reciprocating, particularly related to eye contact.

Some respondents bristled when asked to comment on eye contact. One argued with the notion altogether, writing, "Eye contact?!? You mean contact with my camera? This isn't even possible. But I do look for their facial expressions in response to what I say." And this respondent was not alone. "Due to how cameras work," wrote another, "Eye contact is group or nothing. That is to say that I'm either giving eye contact or I am not, for the entire class. I do not focus too much on getting eye contact from students." This discussion of eye contact reflects a significant proxemospatial and interface design issue present in the Zoom platform. The interface redefines eye contact, and the definition varies from one person to the next, creating a polarizing dilemma. Another respondent discussed their strategy for eye contact: "I have a little guy taped next to my webcam and I talk to him. Students say it's like I am looking right at them all of the time, which is better than they'd experience if we were all in a physical room." It is interesting though that this last instructor reports mastering the idea of simulating eye contact but in doing so has lost all the reciprocity it offers.

Other respondents tried to balance the simulation of eye contact with the need for reciprocity. Many responses shared the sentiment of this participant: "I try to make eye contact, but I'm not always sure I've mastered the camera mediation. I want to look at their faces and read them, so I'm not always looking onto the camera. But I try to keep it close enough that maybe it works." This attempt to achieve balance was one of the most notorious expressions of this dilemma throughout the data.

Nevertheless, approaches to reciprocating were not solely based on eye contact. One respondent commented on using their face rather than eye contact, explaining their reasoning this way: "I find myself exaggerating facial expressions and hand gestures—nodding vigorously, smiling broadly, using 'thumbs up' frequently. No eye contact, though. I'm looking at the image of the student and as a consequence the student sees me looking away from the camera." From my own experience, the overemphasis of these nonverbal behaviors can create outcomes of engagement and even reciprocity from the others in the Zoom environment. However, others answered this dilemma in an exactly opposite manner, noting, "I have not been able to use as many body gestures or motion this year as I would if I was teaching completely face-to-face. This is because my gestures tend to be more directed to individual students, and on Zoom, you can't necessarily dictate which gestures are intended for whom." These two respondents discuss the same dilemma and the same nonverbal technique but reported diametrically opposed results.

Other techniques to navigate turn-taking included uses of proxemics and artifacts to explore strategies to supplement the lack of eye contact. Discussing proxemics, one respondent noted, "I try to replicate conversational dynamics by adjusting my distancing and use of space; it's more engaging for listeners and feels more authentic." This idea of authenticity often surfaced in the theme of reciprocating. Another participant employed an artifact to create a time delay that encouraged turn-taking, writing,

> I find that when asking a question and waiting for students to respond, it really helps to have a mug to take long sips out of to fill the time. Giving students the space to decide to answer takes a long time, and also a lot of them aren't paying attention. So a moment of silence [plus] the instructor sipping tea and staring into the camera seems to get their attention and helps them realize they are supposed to be answering a question.

Even though this study focused on visual nonverbal cues, other studies have investigated paralanguage and chronemics (the use of time) as sources of Zoom fatigue. In the *Journal of Experimental Psychology,* researchers Boland, Fonesca, Mermelstein, and Williamson (2022) reported on their findings concerning transmission delays in the Zoom platform and their role in creating delays in turn-taking during dialogue. "Just as ballroom dancers must coordinate their moves to the beat of the music," they write, "interlocutors must coordinate their utterances to the rhythm of the conversation" (Boland et al. 2022, 1279). Transmission delays on Zoom, they argue, require more coordination and practice as users become increasingly familiar with the platform. This is a worthy reminder that the participants in the present study pivoted to Zoom suddenly and generally without training or the opportunity for practice.

This dilemma of reciprocating was further indicated by polarized responses in the data, reflecting the struggle that many participants faced here. The tone of the responses indicated unease and uncertainty with this dilemma in comparison to the other three, suggesting that this dilemma represents an uneasy tension between the user and the interface. The camera provides a gateway for interaction, but speaking to the camera resulted in a loss of reciprocity. Likewise, viewing the faces moving in the gallery offered important feedback to the speaker but resulted in a loss of perceived engagement for other listeners and onlookers. The result is a conversation that has the potential to be stilted by the technology, perhaps unless or until users develop a level of comfort with a new pattern of nonverbal interaction.

Self-monitoring

Among the survey responses, the theme of self-monitoring refers to "an instructor's perceptions of watching one's own performance on the Zoom platform while teaching" (McArthur 2021, 8). The comments present in this theme all discuss watching one's own performance either through live monitoring or viewing a Zoom recording of the class. Comments reflected both the recognition of the Zoom interface and its constraints as well as a recognition of the reflection displayed in the Zoom frame.

The self-monitoring aspect on Zoom is the clearest indication that Zoom functions as both a window and a mirror, and in this case a literal mirror. One participant noted, "Sometimes I purposely turn off my self view on Zoom so I'm not so self-conscious of the way I'm looking. Otherwise it can get distracting because I'm inclined to look at myself instead of the students." This and similar reactions to the image of self on screen indicates the dilemma present in the decision to watch oneself or not. This dilemma was often represented as an awareness of personal appearance and body language. For example, another participant commented, "Because I can see my own video I am aware of my expressions more than in an in-person class. I try to make sure I am slightly nodding when they are speaking and have appropriate facial feedback, although I would not say I am always explicitly using a strategy, I just notice my own expressions." Another participant offered, "I have taught a fully Deaf class on Zoom. Whether speaking through an interpreter or signing, I'm always hyper aware of my face."

This hyperfocus on the face was represented in many of the responses, including one participant, who noted, "I don't really have a strategy, but looking at the recordings I noticed that I smile a lot, maybe more than I normally would? It gets a bit lonely for me." Instructors in these cases are seeing themselves as they are or at least in the same frame viewed by students. These comments suggest a discomfort or even unease with the face-forward presentation of the body offered by the Zoom platform.

Participants also lamented their approaches to self-monitoring across other forms of nonverbal communication. One respondent noted, "I hate seeing myself on zoom, so I am not getting any closer to the camera than I have to." This comment on proxemics belies discomfort with the self-monitoring dilemma and the respondent's solution. Additionally, it reminds us about the interaction design represented by the device and interface and their relationship to the user. Another participant offered that instead of a face in a Zoom frame, "I want them to see me as a real human being." Scholars of digital communication often ask whether our digital experiences are any less real than our physical ones. But the point this participant is trying to make is clear. The use of digital augmentation removes us from our typical approach to interpersonal connection and replaces it with something notably different.

The idea represented in the theme self-monitoring is one of the earliest critiques of the Zoom platform. In Bailenson's (2021) explanation of Zoom in *Technology, Mind, and Behavior,* self-monitoring is noted as one of the primary explanations for the construct of Zoom fatigue. "Zoom users," he writes, "are seeing reflections of themselves at a frequency and duration that hasn't been seen before in the history of media and likely the history of people" (11). He notes exceptions for dancers in studios and others who work in halls of mirrors, but his point is well taken. Zoom lets us see ourselves in real time, with varying responses.

In the *International Journal of Eating Disorders,* researchers Jennifer Harriger and Gabrielle Pfund (2022) explored the notion of appearance satisfaction among Zoom users. Their study indicates that time spent video chatting is associated with appearance satisfaction and suggests that this finding counters earlier claims of the opposite, noting that individuals may be habituating to video chatting platforms or becoming more accustomed to seeing themselves on screen. These assertions were also mitigated by the auto-enhance and hide-self functions of the Zoom platform. As they noted, this idea of self-monitoring on Zoom is primed for research as individuals continue to use the platform for all sorts of online interpersonal interactions.

Concluding Concepts

The proxemospatial design of the Zoom interface is an area ready for research, particularly as we become collectively more attuned to its design. The nonverbal dilemmas raised here offer a grounding for this study and invite us to explore the negotiation of these dilemmas across contexts. This is indeed one of the most salient concepts in proxemics that relates to almost all nonverbal communication. Once unhidden, the hidden dimensions of nonverbal interaction become part of our decision-making patterns. In this study, participants even noted this type of revelation. One commented, "I have literally never thought about this." Another took this assertion one step further, writing, "I think I could use more visual cues in my zoom classrooms. Your questions started me thinking about this issue." In these responses, an unintended goal of the study became apparent. Through the opportunity to reflect on their own practices, instructors across grade levels found themselves learning and adapting to complex, digitally mediated negotiations of space. Moreover, they had the time to reflect on their work and their uses of the Zoom interface. Such reflection facilitates deeper understandings of digitally mediated interactions and the design of interfaces that promote these encounters.

The presence of these four nonverbal dilemmas on Zoom is directly tied to the role of proxemics in the Zoom interface. The first two dilemmas, animating and replicating, are central to the proxemopetal design of the Zoom interface insofar as they call participants toward a stronger connection to the digital gathering site and the corporate experience. Through the animating and replicating dilemmas, we see a pronounced focus on the content and experience of the group gathering space within the platform. In this proxemopetal design, the interface functions as a window, allowing the participants to see the content and each other and inviting them to create a collective gathering location together. This focus on group gathering and engagement calls for continued research on the negotiation of proxemopetal design in synchronous videoconferencing platforms.

The latter two dilemmas, reciprocating and self-monitoring, are central to the proxemofugal design of the Zoom interface insofar as they call the participant's attention back to the interface itself and back to the participant's own situation. The focus of the user in these two dilemmas is placed toward the interface and the self. In this way, the Zoom interface calls participants back from the interface into their own spaces and invites them to reflect on their own uses of the digital gathering space. In these two dilemmas, the attention of the user leans back toward the proxemofugal design of the interface as mirror, reflecting back to the user both the platform's interface controls and the user's own self-image.

Across these dilemmas, the proxemic interactions in the Zoom space are further complicated by the visual organization of the space itself. When digital technology brings people from far distances together on a computer screen, they become faced with information that they may not be used to processing— particularly when that information includes dozens of people at once. In a seated classroom setting, all copresent others would not be in our personal space at once. Some would necessarily be engaged at social or even public distance. These others would be dispersed at varying distances from the instructor and each other. But on a Zoom screen, everyone is the presenter and also on the front row. Indeed, the Zoom platform brings all users into the interpersonal distance of personal space together, all at once. According to Hall's (1966) *Chart Showing Interplay of the Distant and Immediate Receptors in Proxemic Perception*, this is characterized by visual encounters like seeing details of the face and eyes, a gaze that includes only the visual proportion of the upper body, the likely use of compact gestures, and hearing clearly the voices of others without the need for amplification (126–7). Notably, Zoom does not provide the thermal, olfactic, or haptic perceptions available to us in encounters in a copresent physical environment. But we must wonder whether Zoom functions as a *trompe l'oeil* for our visual or auditory sensations due to this unique experience of personal distance.

As the commentary surrounding the reciprocating dilemma references, the challenge of reciprocating visual interaction with twenty or more others in any physical environment is amplified in the Zoom platform. We simply cannot give eye contact and make eye contact concurrently. In a physical space, eye contact with twenty others would be progressive in nature. Looking at the Zoom gallery therefore begs us to focus our attention on one or two others in the gallery rather than trying to take in every person all at once. Moreover, each user has the ability to change their vantage point in the platform to view the scene as a gallery or to focus on a single individual user, whether that person is the current speaker, a party of particular interest, or a randomly selected other.

Finally, the use of proxemospatial design to characterize these dilemmas opens the door to new perspectives on the use of nonverbal communication broadly, and proxemics specifically, in digital gathering spaces. Considering these interfaces to be spaces of group connection changes the way that we can discuss the activities therein. If we start to study Zoom rooms from the same proxemic perspective that we might use to study public parks or restaurant seating, we may uncover new and compelling insights that foreground nonverbal communication as a vehicle for stronger, more connective encounters in our digital communities. In these early days of Zoom, these alternate perspectives allow us to explore the interface more deeply than we otherwise might, and to connect our work more fully to the complexity of the encounters we engineer through digital negotiations of nonverbal communication.

Note

1   All future quotations and/or references to respondents' survey answers, unless otherwise noted, refer to McArthur (2021) unpublished data.

References

Bailenson, Jeremy N. 2021. "Nonverbal Overload: A Theoretical Argument for the Causes of Zoom Fatigue." *Technology, Mind, and Behavior* 2, no. 1: 1–6. https://doi.org/10.1037/tmb0000030.

Boland, Julie E., Pedro Fonseca, Ilana Mermelstein, and Myles Williamson. 2022. "Zoom Disrupts the Rhythm of Conversation." *Journal of Experimental Psychology: General* 151 no. 6: 1272–82. https://doi.org/10.1037/xge0001150.

Bolter, Jay David and Diane Gromala. 2003. *Windows and Mirrors: Interaction Design, Digital Art, and the Myth of Transparency.* Cambridge, MA: MIT Press.

Glaser, Barney G. and Anselm L. Strauss. 1967. *The Discovery of Grounded Theory: Strategies for Qualitative Research.* New Brunswick: AldineTransaction Publishers.

Hall, Edward T. 1966. *The Hidden Dimension.* Garden City: Anchor Books.

Hall, Edward T. 1968. "Proxemics [and Comments and Replies]." *Current Anthropology* 9, no. 2/3: 83–108. https://doi.org/10.1086/200975.

Hall, Edward T. 1974. *Handbook for Proxemic Research.* Washington, DC: Society for the Anthropology of Visual Communication.

Harriger, Jennifer A. and Gabrielle N. Pfund. 2022. "Looking Beyond Zoom Fatigue: The Relationship Between Video Chatting and Appearance Satisfaction in Men and Women." *International Journal of Eating Disorders* 55: 923–32. https://doi.org/10.1002/eat.23722.

Haywood, Benjamin K., Diane E. Boyd, and John A. McArthur. 2023. "Purpose, Place, and People: How the Pandemic Helped Foster Open and Inclusive Course Design." *To Improve the Academy: A Journal of Educational Development* 42, no. 1 (Spring): 113–43. https://doi.org/10.3998/tia.1680.

Marras Tate, Joanne C. 2023. "'Hello, Garden Eel Here': Insights from Emerging Humannature Relations at the Aquarium during COVID-19." *Environmental Communication* 17, no. 3: 218–29. https://doi.org/10.1080/17524032.2021.2014923.

McArthur, John A. 2016. *Digital Proxemics: How Technology Shapes the Ways We Move.* New York: Peter Lang Publishers.

McArthur, John A. 2021. "From Classroom to Zoom Room: Exploring Instructor Modifications of Visual Nonverbal Behaviors in Synchronous Online Classrooms." *Communication Teacher* 36, no. 3. 204–15. https://doi.org/10.1080/17404622.2021.1981959.

Palmer, Chris. 2022. "Silence is a Worship Practice Zoom Can't Replicate." *Christian Century*, January 12, 2022. https://www.christiancentury.org/article/features/worship-practice-zoom-can-t-replicate.

Sommer, Robert. 1969. *Personal Space: The Behavioral Basis of Design.* Englewood Cliffs: Prentice-Hall, Inc.

5

Zoom's Performative Window

Affordances and Constraints

Daniel Paul O'Brien, University of Essex

Introduction

Zoom, you chased the day away [. . .]
Then my whole wide world went zoom

—"Zoom" by Fat Larry's Band

The R&B funk artists Fat Larry's Band sang the lyrics to the popular 1980s hit, "Zoom," approximately forty years before the novel coronavirus would change the world, yet parts of the song uncannily foreshadow things that would come. Consequences of self-isolation through the uncertainty of Covid-19 meant that for most people within a digital infrastructure, their whole wide world would become "Zoom" as Zoom Video Communications, Inc, founded by Eric Yuan (Matsuo et al. 2023, 38), quickly established itself as the dominant mode of communication during the bleak periods of life in lockdown. From education to socialization and employment, Zoom became one of the most popular digital tools of the global pandemic (Pollák et al. 2022, 75) and a window out from (and into) the domestic lives of all who used it.

The backdrop of bedrooms, living rooms, and intimate spaces quickly became a familiar aesthetic of pandemic and post-pandemic life. The mediating Zoom window has simultaneously normalized the virtual copresence of multiple spaces and attendees who have arguably become performers through the video communication platform, mindful of their own personal mise-en-scène. Fastidiousness over lighting, background, and facial habits on camera are instances of Zoom performance at the micro level—we all want to give our best performance. But at the macro level, Zoom has become an intriguing space for performative concepts of selfhood, from virtual puppetry in the form of avatars for glitch-free calling (BBC News 2020) to new types of digital performance and spectatorship. The familiarity of Zoom and other video calling windows as a mediating device has dominated television screens (as well as computer screens) through timely programs and films. *Staged* (Evans 2020–2) and *Coastal Elites* (Roach 2020), for example, use video calling as a central mechanic in each narrative, which was first utilized out of necessity but by the final season of *Staged* became a stylistic choice.

This chapter will consider the creative potential of Zoom as an alternative virtual stage and its ability to remediate films, theatre, and filmmaking practices through its cultural omnipresence. Zoom is a form of remediation that, according to Jay Bolter and Richard Grusin (1999), can be understood as something that "restore[s] [older forms of media] to health" (59) and "can also imply reform in a social or political sense" (60). Remediation, according to Bolter and Grusin (1999), moves a medium closer toward immediacy. Photography is more immediate than painting, film more immediate than photography, television more immediate than film, and Zoom more immediate than all of the above. This chapter considers the remediation of film and theatre through Zoom affordances that aspire to bring closeness through the platform's capacity as an immediate and intimate medium, yet simultaneously presents a distancing virtual window characterized by temporal abruptness, transient intimacy, and the mental and embodied experience that has come to be known as Zoom fatigue.

The paradoxical affordances and constraints of Zoom are considered in this chapter along with its remediating effect on film via Paul Thomas Anderson's *Magnolia* (1999) and Richard Linklater's *Tape* (2001), which are both remediated through Zoom as a video essay and a Zoom show. The structure of this chapter begins with video essays before moving into digital theatre shows and concluding with a documentary film made via Zoom to explore the remediation, affordances, and constraints of the video call software that have become pervasive in cultural consciousness.

The first section of the chapter considers two video essays by Ian Garwood: *Magnolia Zoomed* (Garwood 2020a) and *Slap That Bass Zoomed* (Garwood 2020b), both of which were included in the British Film Institute (BFI) poll for best video essay of 2020. Each video considers how the aesthetic of Zoom can be used to promote a space for inclusivity, which the latter considers through race. The second section of this chapter switches format to analyze Zoom as a theatrical performance space and considers Laurence Owen's (2021) autobiographical solo show, *Some Other Mirror*. Owen's show uses Zoom to reflect upon the pressure of life in lockdown while dramatizing the performer's initial journey in accepting himself as a trans man. Other versions of Owen fuel this battle by showing up on Zoom as conflicting voices in the guise of video callers. This section includes material from an original, unpublished interview with Owen, which I conducted on March 31, 2022.

Finally, this chapter will discuss a documentary created during the second lockdown in 2021, centering on four young people who at the time were preparing themselves to return to school. *WE'RE IN THIS TOO*, produced and directed by Nic Blower (2021), is a unique film not just in terms of its subject matter but also in the manner that the film was made and directed, which was primarily through Zoom. Made during the height of social distance restrictions, Blower was unable to be in the physical vicinity of the four case studies on screen but chose instead to direct the film through Zoom to give an authentic glimpse into the four worlds of each screen persona. This section also contains material from an original, unpublished interview with Blower, conducted on March 30, 2023, about the process of making the film. Throughout each case study, remediated by Zoom, a paradox of the medium's perceived intimate closeness and simultaneous distancing and constraints will be considered.

Magnolia Zoomed

Video essays or videographic scholarship is an alternative multimedia way to approach and analyze screen material by using the tools of an audiovisual format to put forward a new idea. Ian Garwood's *Magnolia Zoomed* is a case in point, but before coming onto this, I will offer some background on the source material of the film.

Paul Thomas Anderson's multistrand narrative, *Magnolia*, is a film set in San Fernando Valley, Los Angeles, which presents the intersecting lives of nine specific characters who are connected by coincidence and chance over the course of a particular day. At over three hours long, the film is considered an epic psychological drama that magnifies a range of personal sufferings across a mosaic structure of interconnected characters. In Garwood's (2015) book, *The Sense of Film Narration*, the author and video essayist comments that "*Magnolia* features a very unusual moment (at least in a non-musical, feature-length, fiction film) in which all the major characters, despite being in separate locations, sing along to the song, 'Wise Up' by Aimee Mann, which can be heard playing non-diegetically throughout the sequence" (128). This transpires as an unexpected montage two-thirds of the way into the film as Mann's off-screen voice leads each character into a specific verse of the song. As Garwood (2015) notes, "[t]he 'Wise Up' sequence presents an ordered interaction between Mann's voice-over and the onscreen voices of the characters who find themselves singing along with her" (135). Garwood (2015) further notes that Mann's leading voice is responsible for numerous parts of the film's narrative. He highlights how "P.T. Anderson claims that various narrative situations in the film were inspired by Mann's songs which pre-existed the writing of the script" (128–9). This reading suggests Mann's significance to the film as a whole, as well as her prominence within the montage sequence, yet she doesn't share the screen in the same way the other characters do.

Garwood's video essay, *Magnolia Zoomed*, created five years after his book in the midst of lockdown life, recreates this montage but does so through the stylization of a Zoom video call between these characters. What Garwood begins as an experimental piece to mark Zoom's growing dominance over communication during lockdown unintentionally (at the time at least) becomes an exploration of the affordances of Zoom as a space to create visibility and provide a virtual platform of togetherness.

Garwood aesthetically utilizes the familiarity of the Zoom meeting room with the screen divided into nine boxes for the nine on-screen characters (two of whom share the same space) and Mann. This aesthetic choice utilizes a function of the video essay known as "multi-screen composition" (Keathley, Mittell, and Grant 2019), which is the use of frames within or beside other frames, or a multiscreen approach. Each box shows the written name(s) of each character/singer in the familiar Zoom font and color scheme of white on black. As each character (in specific order) duets with Mann, a green border around their box lights up, replacing a mute symbol. After each duet, a video image from the film replaces the text, creating the sense of the characters joining a conversation in a Zoom meeting as they sing their duetting verse. Garwood (2021) discusses this rationale in a separate piece of videographic criticism, titled *The Scholarly Video Essay*, where he asserts the following:

> Last year, in response to video conferencing becoming such a central aspect to my life during the pandemic, I made a couple of videos that converted film sequences into a Zoom format. *Magnolia Zoomed* reimagined the "Wise Up"

sequence as an ideal Zoom call, where there are no awkward pauses, everyone takes their turn to speak, and no one speaks over someone else. Apart from reformatting the scene, I also added Aimee Mann as a bodily presence, whereas in the film she is just a disembodied voice.

Journalist and video essayist Leigh Singer has praised Garwood's work for capturing the spirit of life in lockdown mediated through the video call aesthetic. In his contribution to the 2020 *Sight and Sound* Video Essay Poll, Singer claims that *Magnolia Zoomed* could be "2020's video essay anthem" (Avissar et al. 2020). Garwood (2021), reflecting on *Magnolia Zoomed,* has stated that when making it, he wasn't approaching the work as a piece of criticism. As he explains:

> It definitely doesn't lay out an argument explicitly, and it doesn't try to explain its critical perspective in the way my writing on the film does. But if I were to defend the video as an act of criticism [. . .] the key would be to think about the effect of visualizing Aimee Mann. It's an intervention I have made that brings to the surface a quality of the scene that is kept out of sight (if not sound) in the original. Namely that Aimee Mann is as much a character in the scene as the characters we see onscreen.

What Garwood indicates is a sense of inclusivity through the audiovisual mock-up of the Zoom space, where characters and nondiegetic performers (i.e., those who do not usually occupy the same space in a film) can unite and become virtually copresent with one another. This is indeed the essence of what Zoom afforded for many in the real world: the ability to exist with one another in an audiovisual way, when no other means was possible. In *Magnolia Zoomed,* Mann is afforded a platform to be seen as well as heard, as the Zoom call grants her visibility and a sense of copresence between her and the viewer, as well as the other characters. Garwood's videographic film therefore suggests a political framework for Zoom's affordances as a medium—a concept he develops further for his second piece, *Slap That Bass Zoomed,* which enables similar visibility for African American performers (through a Zoom aesthetic) who were denied the spotlight within early Hollywood films.

Slap That Bass Zoomed

Garwood's (2020b) *Slap That Bass Zoomed* focuses upon Mark Sandrich's 1937 film *Shall We Dance* starring Fred Astaire. During a famous scene, Astaire adeptly performs a musical number in a ship's engine room to a group of

African American workers who sing and produce the musical accompaniment of the song "Slap That Bass" (1937), which coincidentally features the lyric "Zoom Zoom Zoom" at several points in the verses. Like the previous example, Garwood makes an intervention in this sequence through a mock-up Zoom call, in which he removes the limelight from Astaire and turns him into a Zoom spectator watching the African American performers, who in the original film are left in the background with attention upon the dancing star.

In Garwood's version, a multiscreen composition in the style of Zoom windows is again utilized to depict an array of Black performers (with names present) from early musical cinema who Garwood notes were similarly denied the Hollywood spotlight. The song shifts to a version performed by Ella Fitzgerald as a montage of performers, including Eunice Wilson, Adelaide Hall, and The Sepia Steppers, are given a visual platform to perform through a range of Zoom windows, as Astaire continues to look on. Garwood's use of the multiscreen composition enables a timeline of visibility across the screen, creating a contemporary visual foregrounding from historical marginalization. This continues up to the present day, in which Garwood also includes a video of Jalaiah Harmon's TikTok "Renegade" dance video. The Renegade dance was created by a fourteen-year-old Harmon in 2019 but controversially went onto become viral after Charli D'Amelio, a more well-known internet personality through algorithmic bias, performed Harmon's dance without crediting the original source. Consequently, the popularity of the viral dance positioned the original Black artist (Harmon) into the background while focusing the spotlight onto the white performer, D'Amelio. This has since been addressed, and D'Amelio now gives dance credits to routines that she performs (Boffone 2022, 21) but it still highlights an injustice for Black artists of screen media. Trevor Boffone (2022) argues that TikTok's algorithms are a "tool of white supremacy" (22) and asserts that "Tik-Tok fame mirrors general fame in the United States, which privileges whiteness and conventional Western beauty standards" (Boffone 2022, 19). Boffone states that although

> Harmon's Renegade growth launched her career, her social and cultural mobility paled in comparison to D'Amelio's lucrative deals and march toward 136 million followers (and growing). Systems of white supremacy give Black teens such as Jalaiah Harmon just enough opportunity so that the system can say that the system doesn't exist. (Boffone 2022)

These contemporary injustices for Black screen artists are mirrored in the origins of early cinema. This is historically evident in the marginalization of

Oscar Micheaux (an early African American filmmaker) in comparison to the controversial D. W. Griffith. While many of Micheaux's films have been lost, the works of Griffith were preserved. *The Birth of a Nation* (1915), which contextually promotes racist ideologies through white supremacy, problematically "heroizes" the Ku Klux Klan. Micheaux's *Within Our Gates* (1920) on the other hand, is a lesser-known film, which challenges Griffith's racism and was subsequently censored (Siomopoulos 2006).

Slap That Bass Zoomed highlights historic and contemporary racial injustices, which Garwood created shortly after the severe and devastating murder of George Floyd, which brought the Black Lives Matter movement into global awareness during the height of the pandemic in 2020. Garwood's Zoom aesthetic film is impactful, like others that came out in response to this tragedy. However, the affordances that Garwood's film puts forward (much like the Zoom platform itself) are curtailed by its fleeting abruptness as the five-minute video concludes with an end meeting button that is clicked, reminding viewers that such affordances of Zoom are ephemeral. Garwood's video essay, as a simulated Zoom meeting, operates as a political performance space where Black musicians, singers, and dancers are given a platform to be seen, heard, and centrally staged, only until the meeting comes to an end and we are ejected back into the unjustness of reality.

Digital Theatre

Digital theatre is constantly evolving and can be traced back to the 1960s and 1970s when developments in computer technology began to be incorporated into theatrical performances. Theatre companies like the Wooster Group, for example, formed in 1975, were pioneers that used audiovisual media and interactive video art in their shows. As Matthew Causey (2007) notes, "[t]he Wooster Group's dramaturgical and performative strategies of appropriation and collage have more in common with video editing, both magnetic and digital, than with traditional theatre practice" (40). Other interactive art groups such as Blast Theory, founded in 1991, have similarly mixed theatrical experience with digital apparatuses to deliver interactive experiences, something that I have discussed in detail in "The Pervasive and the Digital" (O'Brien 2017). Other examples of this interplay between digital technologies and theatrical performance include the use of wearable devices (O'Brien 2020), VR apparatuses (Jarvis 2019), and digital screens (Masura 2020). Since the impact of Covid-19, digital theatre has splintered into a range of other pixelated spaces.

In Aleksandar Dundjerović's (2023) recent book, *Live Digital Theatre*, the author and theatre director discusses the move to a computerized space during the restrictions of lockdown. As Dundjerović (2023) states, "[l]ive Zoom theatre can be seen as a hyperreal that offers an experience of liveness, crossing theatre, computer interactive technology and video/film." Dundjerović (2023) develops this through the concept of "phygital" performance. Phygital is a term that bridges the physical with the digital world to create interactive and immersive spaces for users/audiences. Dundjerović (2023) refers to this as something that began primarily as a rehearsal space but later transcended to a new type of live theatre experience to be hosted across a range of platforms, which includes Zoom. Dundjerović (2023) asserts that "the art of theatre is a live experience for the audience," but digital affordances of live communication and transmission tools have expanded the theatrical stage:

> The same principle of live communication exchanges through digital platforms relate to the live digital theatre, where performers on screen, as a virtual stage space, communicate with the audience, seated not in rows in the auditorium, but at home. In online live productions, the audience engages with the performers through the computer screen, visual projection or television, as they would, in any live interaction on platforms, such as Zoom or Teams. (Dundjerović 2023)

Dundjerović (2023) draws upon Philip Auslander's concept of liveness to back up these points. According to Auslander (2002), liveness is the effect of mediatization, meaning that liveness can only be considered through the binary opposite of something that is recorded. As Auslander (2002) states:

> Historically, the live is actually an effect of mediatization, not the other way around. It was the development of recording technologies that made it possible to perceive existing representations as "live." Prior to the advent of those technologies (e.g., sound recording and motion pictures), there was no such thing as "live" performance, for that category has meaning only in relation to an opposing possibility. (7)

With this in mind, I now turn to a "phygital" Zoom show that incorporates both live and recorded elements to present a powerful performance about gender identity, mediated over the affordances of the Zoom call.

Some Other Mirror

Some Other Mirror is a solo autobiographical show, written, produced, and performed by Laurence Owen (2021). It explores, over the course of forty-five minutes, Owen's deeply personal experience of accepting himself as a transgender man, a process that partly took place during the pandemic, during which Owen began testosterone. In 2022 the performance was readapted for the stage and has been performed at a number of events including the Edinburgh Fringe, but ultimately it began as a "lockdown Zoom show." In 2021 Owen used the digital platform and "pushed Zoom to its limits" to deliver what he describes as a "trans coming-out story" (Laurence Owen, interview, March 31, 2022). The Zoom show of *Some Other Mirror* (performed and recorded in front of a live audience on 29 May 2021) features a number of monologues from Owen on Zoom within the intimate backdrop of his loft bedroom space. The sloping loft windows shower light into the space, illuminating a partially seen bed, clothes rack, and bookshelf in the background. The familiarity of this domestic mise-en-scène frames Owen who is seated, positioned in the middle. Owen welcomes his audience into the Zoom seminar space, while casually commenting upon the music that plays in the background, which sets a familiar and hybridized tone between entertainer and educator, a combination recognizable to anyone who undertook a class during the pandemic.

Through Zoom, Owen delivers a range of dual performances at different levels. His opening greetings toward the audience as they arrive at the Zoom webinar space is natural and unrehearsed, after which Owen fluently crosses the threshold into the reflective and rehearsed monologues about life in lockdown and his own struggles with a gender identity crisis. The specific functionality of Zoom is then proficiently used to stage a series of calls between the live Owen and previously recorded alternative versions, giving the effect of the performer carrying out a conversation with himself in real time. The live Owen is visited by two versions of himself, one as a figure of confidence who supports his decision and the second as a reactionary paranoia who tries to undermine Owen while he is on the precipice of change. The debate between these conflicting voices within the confined Zoom space becomes the subject of this phygital performance.

In 2022, I had the opportunity to meet and interview Owen, who I met in a digital theatre workshop, about using Zoom as a performance tool. In our interview, Owen highlighted that *Some Other Mirror* is predominantly a pandemic show; as he has elsewhere noted: "COVID has always been a part of the show [and] lockdown is an important part of the play's production and story, although it's only explicitly mentioned twice" (Owen 2022). Owen uses songs as a babble of conflicting voices that are both skillfully sung and synthesized to illustrate his internal state. Mixing two strands of temporality to portray mental conflict, Owen is presented singing lyrics from Elton John's (1972) "Rocket Man" in real time upon the Zoom call, which is gradually interrupted by another off-screen version of his voice, crooning lyrics from The Raconteurs' (2006) hit "Steady, as She Goes." As this occurs, the live on-screen Owen visibly reacts by endeavoring to drown out (the off-screen recorded voice) by singing and focusing harder on the Elton John hit. The gender-conscious song titles serve as a symbolic indicator, or road map, of the internal voices that Owen has dealt and battled with. The result affords a snapshot into the perturbed headspace of gender identity, exacerbated through the isolating pressures of life in lockdown.

This sing-off battle ushers in the format of the show, which is fundamentally two versions of the same person (one recorded and one live) appearing to converse in real time, which Owen achieved in the following way:

> I had two copies of Zoom running from my computer at the same time. I had one copy that was joined to the call (with the audience). For the other, I used a virtual camera (a tool in which you can put a screen grab into a camera). With a virtual camera you can basically make/pretend that you have got a camera running and then have your computer put whatever you like into it. For example, you can insert a screengrab or recorded video in the virtual camera, which is what I did. (Laurence Owen, interview, March 31, 2022)

This effect comes to fruition with a recording of a grainy, distorted image of Owen in a Zoom window (contrasted by the much crisper, live version). The recorded image slowly moves backward, revealing the graininess to be from the close proximity of a camera that is looking directly *at* the screen of Owen's desktop computer (juxtaposed with the live Zoom window, which is looking at Owen *through* his computer). The recorded version pulls back further, revealing the back of Owen sitting at his desk with his face displayed on two monitors. As the camera continues to move back, showing more of the space, Owen slowly swivels round to face the camera placed behind him, revealing his private desk space, an angle normally unseen by recipients of a Zoom call. As recorded Owen turns, so live Owen does the same, mirroring the movements and turning away from the live camera as recorded Owen turns to face the camera behind him, creating a strange digital *mise en abyme*. As Owen asserts:

I set up the show in a particular way to give a sense of voyeurism, which is a type of intimacy that is less easy to achieve in an auditorium. When you are in an auditorium you know you are watching a show, whereas if you are on a Zoom call there is still a sense that you are experiencing a meeting, and this is something that is not a meeting. It is a bit unusual, and I try and establish that this is not a typical Zoom call quite early on. (Laurence Owen, interview, March 31, 2022)

As Owen highlights, Zoom affords a specific level of intimacy, which traditional theatre spaces may struggle to achieve. This is primarily due to the structure of an auditorium in which facial close-ups of performers are sparse. This is not the case in *Some Other Mirror*, which utilizes a range of close-up talking heads conversing on a video call, a process that incorporates the live version of Owen (the performer) responding in real time to conversations with his prerecorded self who appears on Zoom to give the illusion of continuous conversation. This is carefully constructed by Owen (the technician) who is in the dual role of performer and stagehand, reinforcing the concept of duality to which *Some Other Mirror* brings to light.

This process of digital intimacy and split subjectivity through media devices is something that I have recently explored in the chapter "Digital Love: Through the Screen/of the Screen" (O'Brien 2023). Within this work I draw on a range of authors, including N. Katherine Hayles and Brian Rotman, to argue how media devices have significantly contributed toward the pluralization of self. Hayles (2008) is one of the early thinkers to highlight how "communicating by email or participating in a text-based MUD (multi-user dungeon) already problematizes thinking of the body as a self- evident physicality" (27). Rotman (2008) likewise claims that digital media technologies change users into parallel forms of self, whereupon their electronic virtual presence comes to exist alongside their organic flesh body. This is a concept that Owen explicitly puts forward through the debates he has with previous versions of himself through the aesthetics of the video call.

At the time of recording that show I was only about one week on testosterone, and at this point I am now nearly a year on testosterone. So, if I continue to do the show, that's an aspect that I hope will continue feeding back in; how much my physical body has actually changed and how that affects the question of identity within the show. (Laurence Owen, interview, March 31, 2022)

In between the time of our interview and this writing, Owen has since adapted and performed *Some Other Mirror* into a stage show at the Edinburgh Fringe

Festival through the theatre company Chronic Insanity. However, Zoom remains a significant platform to Owen who discussed the importance of other Zoom-based shows during the pandemic, in particular, Neal Davidson's adaption of Richard Linklater's (2001) camcorder film *Tape*, which was reimagined and remediated as a Zoom show. The original film takes place in a motel and depicts three reunited friends before an awful truth emerges between the characters, which is caught on a hidden recording device. The Zoom show stages this as an online call but also utilizes each actor's actual living space into the performance. As Owen notes, each actor had their domestic space of their home set up as their character would (Laurence Owen, interview, March 31, 2022). Moments in the show portray each character picking up their device and walking around their room or into a different part of their home to set up a shot. This affords real-time cameras on all three characters throughout the performance. Owen notes interesting moments of staging to guide the spectator on where to look (Laurence Owen, interview, March 31, 2022). For example, if a character is not getting through properly, they are positioned further back in the frame. Or if characters become really intense with each other, they move closer to the camera.

Davidson (2020) has commented directly about the "dramatic betweenness" that Zoom can offer as a new theatrical experience, devising a rubric for actors working with Zoom. This includes getting to know the boundaries of the Zoom frame (like the stage itself) but also being aware of the duality of the camera lens. Like Owen's show, which is dealing with a duality of gender identity, acting for Zoom is a dual process in which "the lens is both your screen partner and your audience" (Davidson 2020). Part of Davidson's structure is for actors to look straight into the camera lens and avoid looking down at other on-screen actors. As Davidson's (2020) rubric proposes, performers are required to "act directly into the camera [to] make eye contact with your partner and audience simultaneously." As Davidson (2020) asserts, this is "how betweenness happens. If all actors look at the camera, the audience sees them having a conversation."

This concept complicates Walter Benjamin's (2008) well-known observation that in film, an "audience's identification with the actor is really an identification with the camera." Benjamin makes this point to highlight the difference between a stage actor and a film actor. The stage actor performs to a live audience and is in a position to adjust their performance from the energy of their collective engagement. The film actor, however, performs to a camera, which stands in for an absent audience. In the remediated case of Zoom theatre, these two strands become entwined as actors perform to cameras obscuring a real-time audience they cannot see or interact with. The audience's liveness and immediacy are afforded by Zoom but are yet simultaneously distanced by Zoom's barriers, which restrict the collective engagement between actor and audience as well as the audience members from one to another. This fragmented and isolating approach to performing a conversation in *Tape* between the cast of three actors is reflected within *Some Other Mirror*. Owen adopts a similar methodology but uses the illusion of a Zoom conversation to perform isolation and eventual mastery over self-doubt and gender anxiety to become a content and confident man. Thus, the affordances of Zoom bring the audience closely into this intimate conversational space, but the restrictions of the seminar function limit viewers from engagement with Owen (and vice versa) as well as constraining collective copresence between other spectators.

WE'RE IN THIS TOO

Davidson's rubric and Zoom direction, combined with constrained copresence, brings the chapter to its final case study. Nic Blower's (2021) *WE'RE IN THIS TOO* is a pandemic film that considers life in lockdown for four teenagers on the cusp of returning to school. Due to strict Covid-19 regulations at the height of the pandemic, Blower directed the film and participants (empowering them each to film and create their own story) directly from within the confines of Zoom. In keeping with the previous case studies considered so far, Blower's (2021) film makes use of Zoom's affordances to give voice to a different marginalized group in the form of young people, who during the pandemic were often left feeling overlooked and unheard. The film, which takes place over two and a half weeks, focuses upon four teenagers (aged between fourteen and fifteen) who are mentally preparing for a return to face-to-face learning. In an interview I conducted with Blower on March 30, 2023, he states that each of the four case studies "felt that they hadn't been listened to. This particular time was a time when adults were doing the talking and making all the decisions and no one was listening to them. So, this was their chance to be heard."

As Blower discussed, the constraints of making a documentary film in the middle of a pandemic was unlike anything he had encountered before, primarily because of the minimal contact he had with them outside of Zoom:

> Everything was done on Zoom. Which is very hard when you're making a documentary because you have no real sense of them or their space and what

their lives are like. Normally when you visit someone you go and visit them at their home, or you see them in a kind of familiar environment where you can build trust and empathy for one another. But that's very hard to do on Zoom, where all you can see is the static frame, without knowing what's going on in the background. (Nic Blower, interview, March 30, 2023)

In our interview, Blower highlighted that the subjects themselves undertook the filming without any prior experience. This had to be taught to them through the limitations of Zoom. The finished effects are professional strands of domestic reality, woven together to show overlapping concerns among the young people, which include: the anxiety of returning to public life, frustration at news reports stating that children are to return to school first, being a lower-risk category than others, along with discussions about the deep mental scars that lockdown has left. Much of the testimony from each subject are answers to questions prompted by Blower through a complex filming structure upon Zoom (which I will return to shortly). As the director notes,

> The construction of the film was set up, so they were filming everything in the run up to going back to school having been away for months. My questions were all really about their mental preparations and how those changed from day to day, as sometimes they felt good about it and other days, less so. There were feelings of fragility of friendships which were amplified due to loss of contact outside of online communication. So, there was a lot of anxiety about whether they would fit in, be recognized, or understood. This was a forced absence at the most critical time of their social development. (Nic Blower, interview, March 30, 2023)

As Blower further notes, this was particularly hard for one subject who was an only child: "she had wonderful parents but was on her own without any siblings to talk to" (Nic Blower, interview, March 30, 2023). This is juxtaposed with another teen who was part of a much larger family but who struggles to find space in the house because everyone is always there. This particular subject is often filmed in the garage, seeking solitude. When shots are in the house, siblings are often part of the background, indicating the consistent activity of the busy environment.

As mentioned, the process for directing the film and the prompts for the testimonies come through an elaborate directing methodology, which was predominantly confined to Zoom:

> We used quite an interesting technique where we asked questions via Zoom but got the subjects to record their answers on the cameras that we had supplied to them. This is because the quality of Zoom was poorer in contrast to the film

equipment. You may notice in these moments that the subjects (who are always looking at the Zoom camera) are not always looking directly at the film camera. Instead, they may be looking down, past or to one side of the screen as they respond to the questions via Zoom because sometimes, they may not manage to get the camera position right next to the Zoom [computer] camera. (Nic Blower, interview, March 30, 2023)

As the filmmaker highlights, this made the recording process more elongated than usual because he was not able to check footage on the spot. Instead, Blower was having to safely collect the memory cards from each subject (while supplying new ones) and check the footage from his own home, before giving any feedback or notes via Zoom. As he discussed, often his direction to subjects through this process was to take more time and slow things down. "There was an initial tendency in the early footage for everyone to film quite quickly because that seemed to be the feeling they were having: pent-up frustration. But we managed to slow it down, and gradually the film became more mediated and measured" (Nic Blower, interview, March 30, 2023).

This protracted process was also reflected in the post-production phase, as the film was also edited remotely through discussions over Zoom. As Blower highlights, he was having to deliver the memory cards from his subjects to his editor who was cutting the film and then sharing the edited scenes back to Blower via Zoom, where lag time became the new obstacle (Nic Blower, interview, March 30, 2023). When sharing content over Zoom, audio can often be heard before the picture can catch up, creating a disorientating effect. But the perseverance of the editor, director, and filmed subjects pays off with a unique film (structured by Zoom) that affords a two-week window into the lives of young people who are given a platform to speak out on behalf of an overlooked age group.

WE'RE IN THIS TOO is significantly underpinned by the affordances of Zoom, which at times are aesthetically indicated by structuring all four case studies on screen simultaneously, as if they were in a Zoom meeting with one another. This, however, was never the case, and the four young people did not meet during the production process. The small Zoom-style windows in moments of this film are symbolic in one of the subject's final thoughts, when she notes how the documentary affords viewers a tiny window into two weeks of her life, where we knew nothing about her before or after. This is perhaps the very essence of Zoom: a format that offers a virtual window to anywhere and anyone but abruptly concludes once the call is ended. However, such virtual windows, as the case studies attest, can change perception dramatically.

Conclusion

This chapter has considered how Zoom can be a performative space through a range of different media practices, aesthetically in the case of videographic work, dramatically through phygital spaces of Zoom theatre, and practically from behind the camera lens of documentary filmmaking. Throughout these disparate methodologies, the concept of Zoom, visually, actually, or practically, is central to each experience. Across the different formats Zoom has been portrayed as not only a performative space but also a space used for inclusivity through its unique affordances that allow people to be seen, heard, and copresent to one another in an audiovisual way. Garwood's video essays capitalize on these affordances, which offer a virtual stage for classic and contemporary African American performers without having to relinquish the spotlight. However, the restrictive and temporary nature of Zoom, which Garwood reminds us of with the end call button, is fleeting and lasts only as long as the song.

The phygital performance of Owen's *Some Other Mirror* affords a different type of awareness and intimacy from the position of gender. Close-ups upon the face and space of Owen's surroundings afford viewers a more personal experience than what they might have at an auditorium. But once again, this Zoom show comes with a set of restrictions that neutralizes such closeness, primarily through the distance between performer and viewer (as well as viewer-to-viewer contact), through the Zoom seminar setting. Blower's *WE'RE IN THIS TOO* completes this pattern by similarly affording a glimpse into a private world, which is little more than an abrupt window that can quickly open and close at any point without introduction or conclusion.

As these case studies indicate, Zoom as a communication tool and performative practice can be used creatively to unmute anyone to counter marginalization and to become seen, heard, and temporally present. But such affordances can be fleeting and restrictive, often ending with a click of a button.

References

Anderson, Paul Thomas, dir. 1999. *Magnolia*. Burbank: New Line Cinema.
Auslander, Philip. 2002. *Liveness: Performance in a Mediatized Culture*. New York: Routledge.

Avissar, Ariel, Cydnii Wilde Harris, and Grace Lee. 2020. "The Best Video Essays of 2020." December 26, 2020. https://www.bfi.org.uk/sight-and-sound/best-video -essays-2020.

BBC News. 2020. "Virtual Puppets Promise Glitch-Free Video Calls." October 10, 2020. https://www.bbc.co.uk/news/av/technology-54482425.

Benjamin, Walter. 2008. *The Work of Art in the Age of Mechanical Reproduction.* London: Penguin Books.

Blower, Nic, dir. 2021. *WE'RE IN THIS TOO.* Colchester: University of Essex & Essex County Council.

Boffone, Trevor. 2022. *TikTok Cultures in the United States.* New York: Routledge.

Bolter, Jay David and Richard Grusin. 1999. *Remediation: Understanding New Media.* Cambridge, MA: MIT Press.

Causey, Matthew. 2007. *Theatre and Performance in Digital Culture: From Simulation to Embeddedness.* New York: Routledge.

Davidson, Neal. 2020. "Using Zoom in a New Way to Create Theatre That Transforms How We See." December 11, 2020. https://dctheaterarts.org/2020/12/11/using-zoom -in-a-new-way-to-create-theater-that-transforms-how-we-see/.

Dundjerović, Aleksandar Sasha. 2023. *Live Digital Theatre: Interdisciplinary Performative Pedagogies.* New York: Routledge.

Evans, Simon, dir. 2020–2022. *Staged.* London: BBC One.

Fat Larry's Band. 1982. "Zoom." Track 3 on *Breakin' Out.* WMOT Records.

Garwood, Ian. 2015. *Sense of Film Narration.* Edinburgh: Edinburgh University Press.

Garwood, Ian. 2020a. "Magnolia Zoomed." https://vimeo.com/420644447.

Garwood, Ian. 2020b. "Slap That Bass Zoomed." https://vimeo.com/430707925.

Garwood, Ian. 2021. "The Scholarly Video Essay." https://vimeo.com/522393987.

Griffith, D.W, dir. 1915. *The Birth of a Nation.* New York: Epoch Producing Co.

Hayles, N. Katherine. 2008. *How We Became Posthuman: Virtual Bodies in Cybernetics, Literature, and Informatics.* Chicago: University of Chicago Press.

The Ink Spots. 1937. "Slap That Bass." By George Gershwin. Decca Records.

Jarvis, Liam. 2019. *Immersive Embodiment: Theatres of Mislocalized Sensation.* Basel: Springer Nature Switzerland AG.

John, Elton. 1972. "Rocket Man." Track 5 on *Honky Château.* DJM.

Linklater, Richard, dir. *Tape.* 2001. Santa Monica: Lionsgate Films.

Masura, Nadja. 2020. *Digital Theatre: The Making and Meaning of Live Mediated Performance, US & UK 1990–2020.* Basel: Springer Nature Switzerland AG.

Matsuo, Tokuro, Takayuki Fujimoto, and Ford Lumban Gaol. 2023. *Innovations in Applied Informatics and Media Engineering.* Basel: Springer Nature Switzerland AG.

Micheaux, Oscar, dir. 1920. *Within Our Gates.* Sioux City, Iowa: Micheaux Film & Book Company.

O'Brien, Daniel. 2017. "The Pervasive and the Digital: Immersive Worlds in Blast Theory's 'A Machine to See With' and Dennis Del Favero's 'Scenario'." *International*

Journal of E-Politics 8, no. 3: 30–41. https://www.igi-global.com/article/the-pervasive
-and-the-digital/186962.

O'Brien, Daniel. 2020. "Extant's Flatland: Disability and Postphenomenological
Narrative." *AModern* 10: Disability Poetics 1–1. https://amodern.net/article/flatland/.

O'Brien, Daniel. 2023. "Digital Love: Love Through the Screen/Of the Screen." In *Love
and the Politics of Intimacy: Bodies, Boundaries, Liberation*, edited by Stanislava
Dikova, Wendy McMahon, and Jordan Savage, 111–26. New York: Bloomsbury
Publishing.

Owen, Laurence. 2022. Interview. "Some Other Mirror." *The New Current*, August 4,
2022. https://www.thenewcurrent.co.uk/some-other-mirror.

Owen, Laurence. 2021. "Some Other Mirror." May 29, 2021. https://www.youtube.com/
watch?v=ONU-EUteyLI.

Pollák, František, Jakub Soviar, and Roman Vavrek. 2022. *Communication Management*.
London: IntechOpen.

The Raconteurs. 2006. "Steady, as She Goes." Track 1 on *Broken Boy Soldiers*. XL
Records.

Roach, Jay, dir. 2020. *Coastal Elites*. New York: HBO Films.

Rotman, Brian. 2008. *Becoming Beside Ourselves: The Alphabet, Ghosts, and Distributed
Human Being*. Durham: Duke University Press.

Sandrich, Mark, dir. 1937. *Shall We Dance*. New York: RKO Pictures.

Siomopoulos, Anna. 2006. "The Birth of a Black Cinema: Race, Reception, and Oscar
Micheaux's Within Our Gates." *The Moving Image: The Journal of the Association of
Moving Image Archivists* 6 (2): 111–18.

6

Eigengrau

Reimagining Videoconferencing as a "Slow Platform"

Craig Fahner, New York University

Introduction

In May of 2020, with Covid-19 lockdown measures in full swing, I, like most of my peers, found myself spending nearly all my time confined to my home, relying on digital platforms to communicate with the people I would ordinarily see in person. On the one hand, it seemed that monopolistic digital platforms had largely seized on the disruption to everyday life wrought by the pandemic. In the absence of the possibility of in-person congregation, platforms happily intervened with videoconferencing technologies and digital workspaces, all the while maintaining their long-held business practices that compromise individuals' privacy and monopolize human attention. If platforms' business models were already on the path toward integrating themselves into all aspects of daily life, then the pandemic seemed to be a moment in which they might truly consolidate their hold on digital sociality. On the other hand, it seemed that the extraordinary circumstances brought about by the pandemic could serve as a moment to shift public imaginaries around digital platforms toward radical forms of togetherness. The urgent need to combat the isolation of lockdowns demanded inventive solutions to fill the void.

My research during this time focused itself on how dominant digital platforms limit the expression of human togetherness and how the artistic production of platforms might activate otherwise unexplored forms of digital sociality. Building off of several historical precedents and contemporary open-source software initiatives, this research leverages artistic deployments of software to create experiential sites where alternative models for digital collaboration, connection, and communication can be encountered. Rather than exclusively proposing entirely novel technical forms, this work instead acknowledges the material possibilities that already exist in open-source programming initiatives and community-oriented activism, aiming to use artistic creation to create opportunities for these alternative models to reach the public. This chapter reviews existing initiatives in this area, drawing them together into a working method for the creation of artworks-as-platforms that examine and counter the dominant logic of widely adopted platforms like Zoom.

Zoom's features are largely organized around notions of productivity and expediency. Even though Zoom is used for many types of social activities, the logic of workplace productivity is nevertheless imposed onto all platform interactions. Key to this logic is the notion of visual engagement: being seen, seeing others, and seeing oneself. Zoom demands a form of self-conscious attention that operates in line with other monopolistic platforms' imperatives to promote constant sharing and watching. As such, the widespread use of Zoom and other videoconferencing platforms results in expanded opportunities for individuals' attention to be appropriated as a commodifiable object. The work presented in this chapter seeks explicitly to invert these dominant values of productivity and visuality through the production of an alternative platform that emphasizes a slower form of engagement and operates by a nonvisual modality. Documenting a period of artistic production in 2020, this chapter presents *Eigengrau*: an experimental videoconferencing platform that asks its users to close their eyes to the commodification of visual attention and instead to listen deeply and contemplate slowly the presence of others at a distance.

Theorizing the Artwork-as-Platform

Bernard Stiegler (2010), in *For a New Critique of Political Economy*, challenges Margaret Thatcher's well-known catch phrase assertion that "there is no alternative" to the market economy with the counterargument: "There are lots of alternatives" (123). The formation of radically different models for building the world is, for Stiegler, entirely within our grasp and is a matter of reclaiming our imagination from the forces of capital, which foreclose upon creativity and collectivism.

Gehl and Synder-Yuly (2016) note that Stiegler's work gestures toward the potential of digital communication networks and social media to bring about alternate economies of contribution (79)—systems based on sharing, collaboration, and emotional expression. They argue that corporate forms of social media such as Google and Facebook, rather than reifying Stiegler's dreams of an "economy of contribution," instead perpetuate extractive economic models, continuing to fragment sociality and undermine our ability to express care to one another (Gehl and Synder-Yuly 2016, 79). Alternative forms of networked digital media, they argue, are necessary to prevent these tendencies from continuing to proliferate into all aspects of sociality.

Zoom, Teams, and Meet, like other dominant digital platforms, benefit from monopolistic and totalizing network effects and have made themselves successful largely by rendering alternative models unappealing or downright invisible (Sevignani 2013, 335). The fact remains, however, that there are many alternate models for digital communication platforms—both historical and contemporary—that embody entirely different values than their more popular monopolistic counterparts. Here, I will outline some of these initiatives, arguing that *there are lots of alternatives*, and that the task at hand for artists, researchers, and activists is to help demonstrate the possibility of alternative models that oppose the digital enclosures of monopolistic digital platforms.

In the age of monopolistic platforms, open-source technologies have emerged that embed vastly different values in their designs, revealing that alternatives are indeed technically viable. Following in the footsteps of early alternative networking sites like RiseUp.net (Gehl 2014, 156–7), such initiatives include alternative protocols that organize participation into distributed and decentralized topologies like IPFS (n.d.) and alternative, open-source social media platforms that, to quote Gehl and Synder-Yuly (2016), "seek to ameliorate the problems posed by Facebook et al." (80).

Gehl (2015) draws a comparison between alternative social media platforms and existing forms of alternative media such as community radio, independent presses, and weekly newspapers, identifying that these precedents share values of citizen empowerment, participatory democracy, and the opposition of hegemonic forms of representational power. While these existing media forms have found points of access and intervention on the FM dial, in street corner newspaper boxes, and in independent bookstores, access to their online equivalents meets significant challenges in breaking through the all-encompassing enclosures of monopolistic media platforms to reach audiences. In 2019, Klawier et al. (2021) conducted a survey of internet users' awareness and consumption of online alternative media. The study found that only 5 percent of surveyed users could identify an alternative media outlet by name. "This suggests," they conclude, "that only a small avant-garde of internet users—showing a high affinity for political content and low trust in mainstream media—is aware of alternative media" (Klawier et al. 2021, 13). The challenge at hand for alternative platforms is to seek meaningful sites of intervention that work to break through the ideological hold of monopolistic platforms' network effects and algorithmic enclosures.

Participatory artworks that take the form of experimental digital communication technologies are well suited to reveal to the public the radical possibilities of alternative communication platforms. Creative methods that prioritize an aesthetic of playfulness and participatory inclusion can effectively carve out a space in the public imagination around how users might claim more agency in the ways that they share, collaborate, and communicate online. A number of media artists have demonstrated how digital platforms can be rethought through participatory experiments. *Smell Dating*, a project by Tega Brain and Sam Lavigne (2016), for instance, reimagines dating apps like Tinder using novel modes of interaction. Where Tinder and similar apps rely primarily on photographs of users' faces, which can be expediently "swiped" to signal approval or disapproval, *Smell Dating* makes use of no images of faces at all. Instead, the app relies on the user's sense of smell. According to the project's website, participants are asked to leave an imprint of their scent on a T-shirt provided by the dating service. The shirt is then mailed to the artists, who, in return, send swatches of different T-shirts worn by other individuals. If the user is particularly attracted to the smell of a given sample, the artists then facilitate the exchange of personal details (Brain and Lavigne 2016).

feral.earth, a web-based project by Austin Wade Smith (2021), similarly reflects a reimagining of digital communication along alternative modalities. *feral.earth* is a website hosted on a custom-built solar-powered server. Certain links on the site become active when different environmental and ecological conditions are met. During rainfall, a link to a video appears; during gusts of wind, instructions for building kites appear (Smith 2021). In creating a site whose behavior is entirely contingent on actual environmental conditions, Smith's server behaves more like an organism than an infallible and invisible unit in a server farm: it is vulnerable, it opens up when nourished by sunlight, and it completely shuts down when it has received no nourishment. *feral.earth* acknowledges that websites are not immaterial phenomena but rather are part of existing ecologies.

Smell Dating's snail-mail pace and *feral.earth's* season-contingent hyperlinks both operate in the spirit of the "slow computing" movement. Coined by columnist Nathan Schneider in 2015 and explored in Kitchin and Fraser's 2020 book of the same name, slow computing, like the related "slow food" movement, refers to mindful online practices that resist monopolistic platforms' ever-accelerating command over user attention (Kitchin and Fraser 2020, 11). For Kitchin and Fraser, slow computing represents an "ethics of digital care—of self-care and of care for others" (11). In pushing back against the pressure imposed by platforms like Zoom, which work to produce attentive and productive subjects, slow approaches to computing can prioritize alternate values—"enjoyment, patience, individual and collective wellbeing, sovereignty, authenticity, responsibility, and sustainability" (11). Artworks can contribute significantly toward injecting the values of slow computing into public consciousness.

Projects like *Smell Dating* and *feral.earth* do not necessarily represent a feasible or practical alternative to mainstream communications technologies. Rather than being built to handle the scale of mass audiences, these projects instead prioritize a speculative approach, in which users might temporarily disengage from the technologies they are habituated to and communicate along radically different rules and values. They serve as counternarratives—or, to borrow McLuhan's (1997) term, *counterenvironments* (224)—that articulate other possible ways of living and communicating, reinvigorating the spirit of contributive interconnectivity at the heart of early internet imaginaries (Flichy 2007, 5; Turner 2006, 38). Artists, in producing alternative platforms, wield a tremendous power to reorient technological futures toward what Vilem Flusser (2011) refers to as their "dialogic" potential (4). Flusser argues that artists who experiment with communication technologies open up the possibility of radically transformed political formations, in which individuals might be released from the drudgery of conventional forms of work and ownership, toward new forms of participation that "constantly [generate] new information and new adventures" (86).

Artworks-as-platforms serve as "micro-utopias" (Bourriaud 2002, 31), which erase and play with the rules that ordinarily govern online sociability. Drawing from Félix Guattari's notion of micropolitics (1984, 83), Nicolas Bourriaud argues that artworks create a relational space in which utopian social configurations can be temporarily staged and encountered. Relational artmaking, in this sense, does not, at its core, produce objects or commodities but instead produces new forms of sociability (Bourriaud 2002, 33). Artists who produce speculative platforms operate from this relational imperative. Replacing the gallery with the web server or communication protocol, artists working with experimental platforms produce new social forms that oppose the commoditization of digital communication. These practices introduce radical modes of communication into circulation so that they may begin to shift the imaginaries that shape the public's expectations around what digital sociability is and what it could be.

Artworks-as-platforms, by rewiring digital sociability, call into question the institutional rigidity of Zoom and other dominant platforms, which appear to the public as incontrovertible default systems to which all users must subscribe. Instead, artistic reconstitution of platforms reveals the malleability of software: it is a negotiable material that can be reshaped to represent preferable values. The playfulness and aesthetic experimentalism made possible by artworks-as-platforms serve as an accessible entry point in which users can encounter these alternate possibilities and preferable values. *Eigengrau*, the piece profiled in the following case study, follows this approach by rewiring the affordances, values, and modalities of the videoconferencing platforms that have become increasingly vital to everyday work and sociality from the Covid-19 pandemic onward.

Initial Research into Alternative Videoconferencing Platforms

I first initiated this research in April of 2020. With the first wave of Covid-19 in full swing, the world remained in lockdown for the foreseeable future. Videoconferencing platforms had stepped in to take the place of not only the workplace and the classroom but also all manner of social gathering spaces. Services like Zoom, Microsoft Teams, and Google Meet became ubiquitous tools for facilitating meetings online. Intending to explore alternative ways of experiencing togetherness at a distance using online tools, I first opted to assess the dominant values and affordances of these major platforms.

Teams, Meet, and Zoom are, first and foremost, productivity platforms designed to be integrated into a workplace infrastructure. Meet and Teams are extensions of Google and Microsoft's respective workplace productivity platforms. Each of these platforms dictates access according to different levels of user privileges. Administrators can schedule meetings and access detailed statistics logs about users' attendance and participation in video chat sessions, contributing to the increasing tendency of employers to monitor workers' productivity using digital tools—especially tools whose data collection operations are far from transparent to end users. The visual affordances of these videoconferencing platforms, too, produce a panoptic sense of being watched, in which each participant could possibly be scrutinized by a superior or by another participant at any moment. Each of these platforms organizes participants into a grid of others, while also mirroring, by default, each individual user's image back toward them, creating a situation in which each participant is simultaneously watching and being watched.

The increasing use of videoconferencing platforms in place of in-person meetings produces new discomforts and vulnerabilities that are otherwise absent in the sphere of physical interactions. Constantly checking in with one's own image produces a disorienting effect. The cognitive burden of this disorientation is commonly referred to as "Zoom fatigue" (Bailenson 2021, 1; Nadler 2020, 2; see also Crano, in this volume)—a sense of exhaustion associated with extended use of videoconferencing platforms. In one study investigating the effects of Zoom fatigue, a respondent noted that they coped with disorienting feelings by dimming their screen; another described looking away from the Zoom interface entirely; others turned off their cameras (Vidolov 2022, 13). These tactics of withdrawal, while perhaps effective in mitigating the disorienting effects of videoconferencing, are ultimately too readily detectable within the panoptic regime of monitoring that is enabled by Zoom and other platforms. Users can be easily arranged into a grid of cells so that they can all be monitored at once for deviations from established behavioral norms. Digital proctoring software combines videoconferencing software with facial recognition technology and has been deployed by universities to detect on-camera gestures that suggest plagiarism or cheating on exams (Beetham et al. 2022, 22), demonstrating the significant vulnerabilities related to facial surveillance enabled by videoconferencing software. Furthermore, a recent forensic analysis of Zoom reveals the extent to which the platform stores user data—chat messages, time stamps, sent and received files, profile pictures, and other such metadata—which could readily be requested by law enforcement agencies or exchanged with other parties (Mahr et al. 2021, 5). These examples demonstrate that videoconferencing platforms are vulnerable to different forms of state and institutional surveillance. Not only do Zoom users watch themselves and watch others, but they may potentially be watched by unseen third parties due to the integration of facial tracking and data collection techniques.

Making Platforms Malleable

Seeking to invert the earlier mentioned features of dominant videoconferencing platforms through a new interactive artwork, I began to examine how the basic premise of videoconferencing could be reimagined through creative coding. Given that videoconferencing was increasingly being used to establish a sense of togetherness while isolated, I wondered what other modalities, values, and affordances might activate forms of togetherness otherwise excluded by platforms like Zoom. As a starting point, I began to investigate the technical backend that facilitates video chatting on Zoom and other platforms. Zoom manages connections between clients using a cloud service that distributes audio and video (Sander et al. 2021, 4). For its free users, Zoom facilitates peer-to-peer connections, in which clients send data directly to one another without the need for a centralized server (Clopper et al. 2020, 14). Google Meet and Microsoft Teams similarly facilitate peer-to-peer connections using the open-source WebRTC protocol, providing centralized servers as a fallback in case peer-to-peer connections are blocked by firewalls (Nisticò et al. 2020, 1). The WebRTC protocol, on its own, does not require any centralized server architecture beyond a simple system that identifies users' respective IP addresses. As I looked further into the WebRTC protocol, it became clear that video chat can easily be accomplished without the massive, centralized business structures represented by Google, Microsoft, and Zoom.

In the fall of 2020, I began to experiment with the creation of my own WebRTC handshaking server. I was able to get a simple peer-to-peer videoconferencing service set up in one afternoon, which allowed users to connect to a video chat session within their respective web browsers. With a simple template established, I experimented with tweaks to the basic videoconferencing paradigm, starting with the positioning of each client's respective video feed. I recalled an awkward scenario that I'd encountered several times during Zoom sessions, in which a user would refer to the person "to their right" or "to their left" in the grid, only to realize that the organization of the grid view is not consistent for each user. This seemed to contradict the spatial organization of participants that is natural to in-person communication. I added a function that would allow each user to position their own video feed anywhere on the screen using a simple drag-and-drop gesture. If one user, for instance, dragged their video to the top right corner of the browser window, it would appear in the top right corner for every other connected user. This resulted in a playful dynamic in which users could reposition themselves several times over the duration of a video session, establishing different spatial configurations.

This experiment demonstrated how readily the dominant paradigm of videoconferencing could be made malleable through creative reconstitution. With little technical expertise and resources, a functional peer-to-peer video chat service could be created that plays with the rigid affordances commonly associated with videoconferencing platforms. With this notion of malleability in mind, I began to wonder how far I could push this paradigm beyond the familiar conventions of Zoom. How else might a sense of togetherness be established in a platform without relying so heavily on this regime of watching? The design of such a platform might necessarily take a radical turn from visuality. Insofar as advertising- and data collection-driven platforms rely so heavily on *watching* as a gesture, a radical alternative would be to do away with watching altogether. Instead, it would ask that users *close their eyes* to the attention economy.

I became fixated on the gesture of closing one's eyes as an act of resistance to the appropriation of user attention by platforms. While computers and mobile devices may allow us to turn off our cameras, they rarely, if ever, ask us to close our eyes or look away from the screen. Doing so would effectively negate the regime of visual affordances that are deployed by platforms to exploit and commodify user attention. Rather than demanding that users' vision always be available for this kind of exploitation, could a platform instead request users to close their eyes as a condition for participation? What kind of alternative modalities of togetherness and attention emerge in the absence of visuality?

Outside of sleeping, the act of closing one's eyes typically signals a deep meditative sense of awareness—one that contrasts starkly with the distracted, fleeting form of attention represented by the cluttered visuality of digital interfaces. It also suggests a sensory shift, in which attention ordinarily occupied by visuality is focused instead on one or more of the other senses. With closed eyes, our sense of the world and ourselves turns away from detached exteriority and toward a more tactile sensibility, in which our relation to what is around us can be more fluidly and imaginatively articulated—not ordered according to a deterministic, visual logic.

Considering how the sensory shift of closing one's eyes might translate into a participatory experience, I was reminded of the work of American composer and musician Pauline Oliveros. Oliveros (2005) is perhaps best known for the concept of "deep listening"—a mode of musical experience in which the listener is attuned not only to the objective qualities of a sound but also to its relation to the listener's body and environment (10). Oliveros distinguishes between *hearing*—the physical perception of sound—and listening, which is "to give attention to what is perceived both acoustically and psychologically" (13). Deep listening suggests breaking through the boundaries prescribed by ordinary hearing and orienting oneself to the dynamic relationship between sounds, the environment within which they are heard, and the human and nonhuman actors who are generating them.

Oliveros (2005) hosted participatory deep listening exercises and sonic meditations, which provided a structure for participants to expand beyond mere "focal attention"—attention limited to a discrete object—and instead toward "global attention," which acknowledges the network of relationships that constitutes the act of listening (25). In some deep listening exercises, participants are encouraged to actively reinforce sounds, producing vocal tones while listening to others. *Tuning Meditation*, for instance, is a group improvisational exercise that asks each participant to simultaneously take a deep breath and then produce a vocal tone while listening to the sounds produced by others (Steinmetz 2019, 127). With each subsequent breath, the individual gradually adjusts the pitch of the sound they are producing, "tuning" themselves to the sound of the overall group. In this exercise, the act of listening orients each person away from their own, isolated voice and instead toward the communal voice. The group of participants tunes itself not to a single, fixed reference point but always in relation to every other individual in the group, finding, through deep listening, a collectively reached point of balance. "The idea of 'tuning' sits next to attunement," writes Julia Steinmetz (2019) on Oliveros's work. "It is an adjustment of pitch, but also of attention. In either matching the tone of your voice to that of another at the furthest reaches of your hearing, or perceiving that another voice has joined your own unique pitch, there is a sensation of attunement, of communication: 'I hear you and I reflect back what I am hearing. I synchronize with you'" (Steinmetz 2019, 128).

Taken as a protocol for communication, Oliveros's *Tuning Meditation* presents a radical shift away from forms of transmission and reception that explicitly individuate each sender and receiver. Rather than transmitting concrete, verbal messages, *Tuning Meditation* transmits a sense of synchrony with others. It is the dissolution of the individual into a shared experience, materialized as a droning, harmonized sound. Through the nonvisual modality of harmonized sound, a rich, dynamic field of sonic representation is established that vividly depicts a moment of synchronous togetherness, "[attuning] the listener to the potential space opened up by shared practices of attention" (Steinmetz 2019, 128).

The busy visual space of advertisements and addictive affordances through which platforms thrive acts as a fusillade of objects of focal attention. A more global form of attention could be fostered by turning—or *tuning*—toward a slower, nonvisual modality. Requesting that users close their eyes as a condition for participation could effectively serve as a cue for users to shut out these diffuse, focal modes of attention and engage with other modalities and other ways of feeling a sense of synchrony with others—at a distance. The act of closing one's eyes, here, functions as an act of refusal to the economies of attention that rely on our persistent vision.

Implementing *Eigengrau* as a Functional Alternative Platform

Drawing inspiration from Oliveros's notion of deep listening, I outlined the features to be implemented in my alternative "videoconferencing" platform:

1. The platform should be able to detect whether a user has closed their eyes.
2. When a user's eyes have closed, an audible tone, unique to the user, should be generated.
3. As other users join the server and close their eyes, their tones should become audible to anyone else connected. Each user's individual tone should combine into a droning harmony.
4. Users should have some simple, gestural means of controlling or tuning the sound they are producing.

With these parameters in mind, I built a simple decentralized broadcast system that bounced data to other users as information was received without retaining any information about individual users centrally. By eliminating any kind of data collection at the protocol layer, this platform already differed greatly from major platforms, for whom centralized data collection is a primary feature. I then implemented a simple audio synthesis system to play back an organ-like tone on user interaction. Each user is assigned a unique tone when they first load the page. The tones are selected from a list of notes within the D minor pentatonic scale—a scale in which each note has a distinct and tuneful harmonic relationship to each other.

With a functional collaborative networked instrument implemented, I began to investigate different methods for tracking users' eyes. I needed to find a way to determine whether a user's eyes were closed, and to be able to track the position of their eyes in relation to the camera, so that a simple means of gestural control over audio timbre and pitch could be established. This functionality demanded the use of facial recognition technology to extract facial geometry from a user's camera feed, allowing certain facial features to be identified and tracked.

While researching options for facial tracking algorithms, I also examined the well-documented vulnerabilities of these systems to race-based exclusions and biases. In one viral video from 2009, a Black man demonstrates that his new Hewlett-Packard webcam cannot detect his face, while its face tracking software has no issue tracking the face of his white coworker (Chen 2009). Ruha Benjamin (2019) connects this incident to a broader history of racial bias in imaging technologies. Benjamin (2019) notes that "photo technologies, both past and present, tend to cater to lighter-skinned subjects" (74). Just as early Kodak film formulas and exposure standards were calibrated to white faces (Benjamin 2019, 70–1), face tracking algorithms have demonstrated a tendency to exclude racialized faces. Inasmuch as the tracking of facial and other bodily gestures relies on the use of photographic technologies, these systems reproduce the tendencies of *epidermalization* (Fanon et al. 2008, 84), with the color and appearance of the face serving as the dominant factor that determines whether an individual is detected. Recent examples have shown that facial tracking technologies that are designed to detect whether an individual's eyes are opened or closed have likewise reproduced such racial biases. In 2016, for instance, a man of Asian descent from New Zealand, using an automated passport approval platform, had his image rejected by a facial tracking algorithm, which insisted that his eyes were closed even though they were not (Ng 2016).

I decided to use a face tracking system based on the Clmtracker library—an open-source facial modeling algorithm based on the MUCT dataset, which was assembled to provide more age, gender, and race diversity than other existing models (Milborrow [2015] 2022). This model provided measurements for the openness of the mouth and eyes. At first, I attempted to create a one-size-fits-all threshold to detect closed eyes. It became apparent after testing this threshold across a variety of faces that such an approach would not work consistently across all users. The distance between the open eyelids is not the same across all individuals, therefore a standardized threshold for eye closedness would necessarily exclude some participants. A calibration system was implemented, allowing each individual user to configure the system to accurately detect their eyes. First prompting the user to click the screen with open eyes, the calibration system takes a reading of the distance between the user's eyelids and then requests a reading while the user's eyes are closed. By creating a system that, rather than automating the detection of the eyes and presuming every face to be alike, instead allowed each user to configure the system to work best for their own eyes, the potential for automated bias was mitigated.

With a reliable eye tracking system in place, audio and visual features were then implemented. Once the user's eyes are detected, they are mirrored back to the user as a live outlined drawing (Figure 6.1). *Eigengrau*'s protocol allowed multiple users to connect to the site and share their own eye coordinates synchronously, with each user's eyes drawn to the screen in a unique color. Rendering eyes to the screen effectively and simply signaled the presence of other synchronous users. Rather than transmitting any identifiable features of individuals' faces, only the coordinates representing the eyes were exchanged, with no identifying data retained by the server. In this sense, this application of facial tracking inverts its typical, surveillant uses: where most facial tracking regimes strive to make the face legible and classifiable as data, this use of facial tracking makes the captured image more anonymous. I implemented several interactive actions by which users' eyes could control the audio synthesis system.

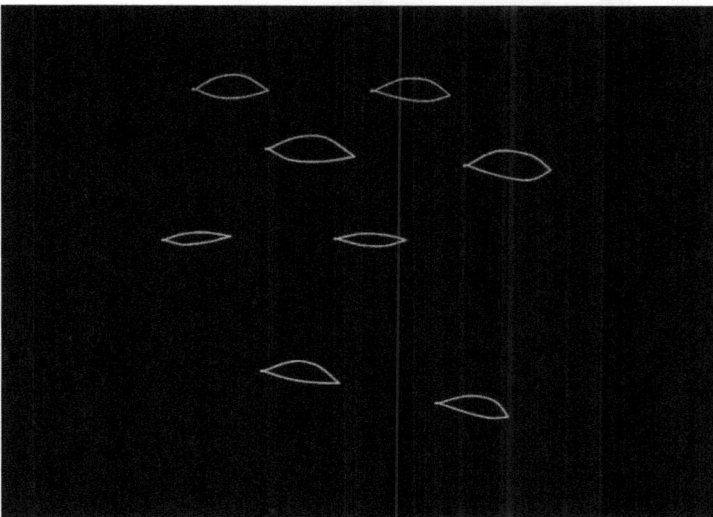

Figure 6.1 Screenshot of *Eigengrau* in use. The eyes of all connected users are displayed as outlines.

When the user moved the position of their eyes, the timbre and pitch of the tone was modulated.

I decided to call the project *Eigengrau*—a term used in the study of optics to describe the almost black color that is seen through closed eyes or in spaces of absolute darkness (Knau and Spillman 1996, 268). *Eigengrau*, as a perceptual phenomenon, is experienced as the residual sensation of lightness in the absence of outside stimuli—as a slow and subtle oscillation of gray light. I appreciated the notion that, even with closed eyes, something is still perceived. *Eigengrau* was a fitting term for the project, as it asks users to close their eyes not to shut out their perception of the world but rather to open up other modalities for experiencing the presence of others. Just as the eyes produce a slower, more subtle variation of perception when closed, this project orients its users toward a more contemplative sense of connection in contrast to the stark and frenetic visual structure ordinarily experienced through digital technologies.

Conclusion: Experiencing *Eigengrau* as a Collaborative Instrument

After testing the platform with peers to ensure that it could withstand ten or more simultaneous connections, I exhibited the completed work at the 2020 OurNetworks conference, which was hosted remotely.[1] The conference, which focused on bringing together activists, scholars, and engineers interested in developing alternative network infrastructures and platforms, was, like many other Covid-19-era online events, laden with back-to-back videoconferencing sessions. While the conference audience had largely adapted to the logic of platforms like Zoom, *Eigengrau* intervened with a starkly different mode of connection, foregrounding slow contemplation in contrast to the visually and cognitively exhausting program of videoconferencing presentations. The launch of the platform at the festival provided an initial opportunity to truly experience the piece with an invested audience.

What I experienced while connecting with others in *Eigengrau* was markedly distinct from other forms of digital communication. I became aware of others not through quantifiable indicators of their identities—profile pictures, screen names, and so on—but through the collaborative production of a sonic object. The object—the shifting, harmonic drone produced through closing our eyes together—was representative of our shared moment of slow, meditative togetherness. As each user closed their eyes, hearing their own sound combined with the sounds of the other participants, there seemed to be a tacit acknowledgment that somewhere, at the other end of the line, there are other people, sharing the same moment of pause, with their eyes closed as well.

The experience rendered each participant not as "dividuals" (Deleuze 1992, 5)—abstract and alienated quantifications of individual bodies—but instead as agents within a rich and dynamic intra-active (Barad 2007, 176) environment that could not be readily reduced to its discrete components. The complex relationship between the tones in a rich, harmonic drone seemed to articulate the sense of dynamic and irreducible entanglement at the heart of Barad's (2007) analysis of complex systems. I was reminded of La Monte Young and Marian Zazeela's 1969 audio installation work Dream *House* (2004). In *Dream House*, four precisely tuned sine wave tones are amplified from the four corners of a room that is washed in a combination of pink and blue lights. Visitors to the space not only hear the dynamic interactions of the tones, but they also hear the way that their own bodies and the bodies of other listeners in the space affect the way those tones reflect and intersect. The total experience is therefore contingent on every listener in the room, each shifting the tones as they orient themselves within the space. *Dream House*, despite offering no words, images, or symbols, nonetheless provides a rich and informative experience for its participants, in which a detailed understanding of the presence of others is articulated through an act of deep listening.

When I first visited the permanent installation of *Dream House* in New York City I stayed in the space for what felt like hours listening to the drone subtly shift as myself and others moved in and out of the space. I found myself attuned to a slow, nuanced sensory modality that contrasted starkly with the frenetic energy of the lower Manhattan streets that lay just on the other side of the door. In a similar manner, I had intended to create a comparable experience with *Eigengrau*—a digital space in which a sense of the presence of others could be established via deep listening, encouraging a slower appreciation of the presence of others articulated through sound. Without explicitly stating this intention, participants seemed to acknowledge that the modal shift facilitated by the project operated as an act of resistance to the attention economy. Technologist Anil Dash, for instance, commented that *Eigengrau* was a calming experience, serving as a "well-earned respite" from the near ceaseless screen time ushered in by the pandemic (Dash 2020).

On its own, *Eigengrau* does not solve the myriad issues common to dominant internet platforms. It does, however, serve as an opportunity for users to disengage from the affordances and enclosures to which they are habituated so that they might experience just how different digital communication might be. As a relational micro-utopia, *Eigengrau* reveals to its participants that communication platforms are malleable things that can be shaped along entirely different values, modalities, and means of access. By virtue of providing users a moment to connect with others outside of the exhausting visuality of everyday digital affordances, it sheds light on the exploitative nature of economies of attention upon which dominant platforms are built. The intention of artworks-as-platforms like *Eigengrau* is not to replace videoconferencing or other communication platforms; rather, they serve to challenge public imaginaries around how sociality expresses itself through networked digital media.

Note

1 *Eigengrau* (Fahner 2020) has remained available online since this initial exhibition and can be experienced at https://eigengrau.glitch.me

References

Bailenson, Jeremy N. 2021. "Nonverbal Overload: A Theoretical Argument for the Causes of Zoom Fatigue." *Technology, Mind, and Behavior* 2 (1). https://doi.org/10.1037/tmb0000030.

Barad, Karen. 2007. *Meeting the Universe Halfway: Quantum Physics and the Entanglement of Matter and Meaning.* Durham: Duke University Press.

Beetham, Helen, Amy Collier, Laura Czerniewicz, Brian Lamb, Yuwei Lin, Jen Ross, Anne-Marie Scott, and Anna Wilson. 2022. "Surveillance Practices, Risks and Responses in the Post Pandemic University." *Digital Culture & Education* 14 (1): 16–37.

Benjamin, Ruha. 2019. *Race after Technology: Abolitionist Tools for the New Jim Code.* Newark: Polity Press. http://ebookcentral.proquest.com/lib/york/detail.action?docID=5820427.

Bourriaud, Nicolas. 2002. *Relational Aesthetics.* Dijon: Les Presses du réel.

Brain, Tega and Sam Lavigne. 2016. *Smell Dating.* https://smell.dating.

Chen, Brian X. 2009. "HP Investigates Claims of 'Racist' Computers." *Wired*, December 22, 2009. https://www.wired.com/2009/12/hp-notebooks-racist/.

Clopper, Alexander J, Eric C Baccei, and Taylan J Sel. 2020. "An Evaluation of Zoom and Microsoft Teams Video Conferencing Software with Network Packet Loss and

Latency." *Worcester Polytechnic Institute.* https://digital.wpi.edu/concern/student
_works/ws859j13x?locale=en.

Dash, Anil (@anildash). 2020. "A Beautiful (and Calming!) Idea, Elegantly
Executed. So Glad to See Art Like This Made on @Glitch. Check It out, It's a
Well-Earned Respite." September 12, 2020. https://twitter.com/anildash/status
/1304585581012439040.

Deleuze, Gilles. 1992. "Postscript on the Societies of Control." *October* 59: 3–7.

Fahner, Craig. 2020. *Eigengrau.* https://eigengrau.glitch.me.

Fanon, Frantz, Ziauddin Sardar, and Homi K. Bhabha. 2008. *Black Skin, White Masks.*
Translated by Charles Lam Markmann. New ed. Get Political. London: Pluto Press.

Flichy, Patrice. 2007. *The Internet Imaginaire.* Cambridge, MA: MIT Press.

Flusser, Vilém. 2011. *Into the Universe of Technical Images.* Minneapolis: University of
Minnesota Press.

Gehl, Robert W. 2014. *Reverse Engineering Social Media: Software, Culture, and Political
Economy in New Media Capitalism.* Philadelphia: Temple University Press.

Gehl, Robert W. 2015. "The Case for Alternative Social Media." *Social Media + Society* 1
(2). https://doi.org/10.1177/2056305115604338.

Gehl, Robert W. 2017. "Alternative Social Media: From Critique to Code." In *The Sage
Handbook of Social Media*, edited by Jean Burgess, Alice E. Marwick, and Thomas
Poell, 330–50. Thousand Oaks: SAGE Inc.

Gehl, Robert W. and Julie Synder-Yuly. 2016. "The Need for Social Media Alternatives."
Democratic Communiqué 27 (1): 78.

Guattari, Félix. 1984. *Molecular Revolution: Psychiatry and Politics.* Harmondsworth,
Middlesex: Peregrine Books; New York: Penguin.

IPFS. n.d. "IPFS Powers the Distributed Web." Accessed July 21, 2021. https://ipfs.io/#why.

Kitchin, Rob and Alistair Fraser. 2020. *Slow Computing: Why We Need Balanced Digital
Lives.* Bristol University Press. https://doi.org/10.46692/9781529211276.

Klawier, Tilman, Fabian Prochazka, and Wolfgang Schweiger. 2021. "Public Knowledge
of Alternative Media in Times of Algorithmically Personalized News." *New Media &
Society*, 25, no. 7: 1648–67. https://doi.org/10.1177/14614448211021071.

Knau, Holger and Lothar Spillman. 1996. "Failure of Brightness and Color Constancy
under Prolonged Ganzfeld Stimulation." In *Human Vision and Electronic Imaging*,
edited by Bernice E. Rogowitz and Jan P. Allebach, 2657:19–29. SPIE / International
Society for Optics and Photonics. https://doi.org/10.1117/12.238731.

Mahr, Andrew, Meghan Cichon, Sophia Mateo, Cinthya Grajeda, and Ibrahim Baggili.
2021. "Zooming into the Pandemic! A Forensic Analysis of the Zoom Application."
Forensic Science International: Digital Investigation 36 (March): 301107. https://doi
.org/10.1016/j.fsidi.2021.301107.

McLuhan, Marshall. 1997. "Address at Vision 65." In *Essential McLuhan*, 219–32.
Milton: Taylor & Francis Group. http://ebookcentral.proquest.com/lib/ryerson/
detail.action?docID=5268426.

Milborrow, Stephen. (2015) 2022. "StephenMilborrow/Muct." https://github.com/
StephenMilborrow/muct.

Nadler, Robby. 2020. "Understanding 'Zoom Fatigue': Theorizing Spatial Dynamics as
Third Skins in Computer-Mediated Communication." *Computers and Composition*
58 (December): 102613. https://doi.org/10.1016/j.compcom.2020.102613.

Ng, Alfred. 2016. "Passport Software Can't Tell This Man's Eyes Are Open." *CNET*,
December 7, 2016. https://www.cnet.com/culture/new-zealand-facial-recognition
-eyes-closed-error-rejected-richard-lee/.

Nisticò, Antonio, Dena Markudova, Martino Trevisan, Michela Meo, and Giovanna
Carofiglio. 2020. "A Comparative Study of RTC Applications." In *2020 IEEE
International Symposium on Multimedia (ISM)*, 1–8. https://doi.org/10.1109/ISM
.2020.00007.

Oliveros, Pauline. 2005. *Deep Listening: A Composer's Sound Practice*. Lincoln, NE:
iUniverse.

Sander, Constantin, Ike Kunze, Klaus Wehrle, and Jan Rüth. 2021. "Video Conferencing
and Flow-Rate Fairness: A First Look at Zoom and the Impact of Flow-Queuing
AQM." *ArXiv:2107.00904 [Cs]* 12671: 3–19. https://doi.org/10.1007/978-3-030
-72582-2_1.

Sevignani, Sebastian. 2013. "Facebook vs. Diaspora: A Critical Study." In *Unlike Us
Reader: Social Media Monopolies and Their Alternative*, edited by Geert Lovink. Inc
Reader 8. Amsterdam: Institute of Network Cultures.

Smith, Austin Wade. 2021. "Feral.Earth." http://feral.earth/.

Smith, Austin Wade (@_newcubes_). 2022. "I've Been Raising Http://Feral.Earth for
a While and Im Happy to Finally Share the Project. A Gentle Rant about Queering
Servers on the Feral Web." March 3, 2022. https://twitter.com/_newcubes_/status
/1499408371971149828.

Steinmetz, Julia. 2019. "In Recognition of Their Desperation : Sonic Relationality and
the Work of Deep Listening." *Studies in Gender and Sexuality* 20 (2): 119–32. https://
doi.org/10.1080/15240657.2019.1594048.

Stiegler, Bernard. 2010. *For a New Critique of Political Economy*. English ed. Cambridge
and Malden, MA: Polity Press.

Turner, Fred. 2006. *From Counterculture to Cyberculture: Stewart Brand, the Whole
Earth Network, and the Rise of Digital Utopianism*. Chicago: University of Chicago
Press.

Vidolov, Simeon. 2022. "Uncovering the Affective Affordances of Videoconference
Technologies." *Information Technology & People* ahead-of-print (ahead-of-print).
https://doi.org/10.1108/ITP-04-2021-0329.

Young, La Monte and Marian Zazeela. 2004. "Dream House." In *Selected Writings*,
10–16. ubuclassics. http://www.ubu.com/historical/young/young_selected.pdf.First
published 1969, Munich: Heiner Friedrich Gallery.

Transverse Networks and the Neoliberal University

The three chapters in this section explore the impact of Zoom on higher education, with attention to the experiences of both faculty and students and the tensions that develop between formal and informal modes of learning and engagement. "Engagement" itself operates as a critical inflection point in these discussions to the extent that, long before Zoom, the adoption of educational technology—from course management systems to classroom "clickers" to gamified learning apps—served to create an equivalence between quantifiable measures of digital engagement and the standards and measures for engaged learning (Kuntsman and Miyake 2022). Zoom, as a widely adopted videoconferencing platform in primary and secondary schools as well as colleges and universities, proved no exception to the rule that tools and instruments that "cut" (Barad 2007) *learning outcomes* and *student engagement* by way of digital measures are well rewarded within corporatized models of higher education.

That is not to say that these same networks of control do not provide potential for modes of interaction that disrupt institutional expectations for efficient, productive digital engagement. Students operating the same interface structures that cede so much surveillance authority to instructors found ways to subvert those tools. It was not uncommon during a period of mandated digital engagement for students to push back against an intrusive instructional technology with cameras off and mics muted, leaving teachers and faculty performing before a grid-framed audience of blacked-out screens. We will expand upon the precarity of these networks, and how this mantra of "digital engagement" serves to obscure that instability, in the final section of this collection. The three chapters in this section, however, offer an explicit focus on the *situation* of faculty and students alike within "Zoom University." They provide an in-depth discussion of the affordances and constraints of digitally mediated learning by way of videoconferencing applications and how these platforms institutionalize a logic of maximum performance within educational settings.

All three of the chapters in this section engage in a form of autoethnography, examining how control networks operate on and through Zoom from the position of graduate student, faculty member, and administrator. Such an approach seems particularly relevant for those of us writing on the topic of a pandemic that is still very much underway, even if we have been made to resume our "normal" lives as if we have passed beyond a crisis stage. It is doubly relevant for those graduate students and faculty members who not only lived through the pandemic but were very much caught up in the changes in teaching modalities and norms for university meetings during stay-at-home orders and social distancing mandates. We would suggest, along with Tomaselli (2015), that autoethnographic methods offer a counterpoint to the positivist approaches to data collection and measurement that underpin the "academic managerialism" that dominates higher education at present. We would also suggest that the Zoom interface itself primes us for an autoethnographic approach, as we find ourselves positioned as both observer and observed in a grid of faces (Crano this volume; Sumner 2024), simultaneously acting as both controlling subject and controlled object. And while this modulation within a larger network of circulation is at the heart of cybernetic control systems (Deleuze 1992), it likewise suggests spaces in which voices otherwise unheard might indeed break through—if not in a sustained way, then as a momentary and precarious irruption, a potential for an "engagement" that does not express itself in terms of performance metrics.

While none of the three chapters in this section takes up an explicit discussion of Guattari's (2015) concept of *transversality*, the notion of dynamic, situated challenges to institutional regimes of authority has considerable relevance to how and where these alternate experiences of engagement may emerge. While the psychoanalytical origins of this term are somewhat irrelevant to the discussion at hand, what is highly applicable is the way in which Guattari uses the image of the transverse line—diagonal, so to speak—as a direction of flow that is neither vertical (top-down, as in traditional institutional power structures) nor a fully undifferentiated (and chaotic), horizontal system. The transverse line offers a particularly resonant metaphor, we suggest, in the context of the Zoom user interface, which presents itself as "a grid of isolated box shapes confined to unbreakable bounds" (Della Ratta 2021). What Guattari describes as a "coefficient of transversality" is, in effect, a measure of to what degree that diagonal "cut" can open potentials that resist falling (vertically) into line. As Genosko (2002) notes, Guattari's transversal marks a moment of virtuality, "a space in which becomings are truly creative—radically open and simply not what is now actual" (75). Transversality marks the scene of "praxic opening" in which set boundaries likewise mark potential relations, unintended headings, and unforeseen challenges to structures of authority and control (Genosko 2002, 108). To what extent, the chapters in this section ask, did Zoom offer a platform for similar moments of praxic opening? While deployed as a tool to "pivot" and "adapt" in the face of a pandemic, and in doing so modulate operation within an ongoing logic of maximum performance, what "minoritarian," haphazard, and temporary openings nevertheless presented themselves "as a potential, creative and created, becoming" (Deleuze and Guattari 1987, 106), be that through backchannel networks, subversive communiqués, or borderland encounters?

In "The Zoom Machinic in Postdigital Learning Ecologies," Kathryn Grushka, Rachel Buchanan, Michael Whittington, and Rory Davis explore the altered learning ecologies of teachers and students brought about by Zoom. Deploying an autoethnographic methodology, and drawing upon Guattari's discussion of machinic assemblages, they explore new formations of human-machine and human-human interaction by way of Zoom that give rise to new potentials for learning—but also new potentials for informatic control. As such, this initial chapter provides a broader theoretical framework for thinking through "encounters" in terms other than those that reduce to a form of digital engagement. The final two chapters in this section focus on the use of tools such as Zoom as a means of subverting traditional power structures in order to create alternate means of communicating, gathering, and connecting outside of academic employers' boundaries. First, Alexis-Carlota Cochrane and Theresa N. Kenney's chapter, "'Networked Togetherness, I Guess ‾_(ツ)_/‾': Subverting the Academic Zoom Chat through the Subcultural Collective," explores how the rapid shift in teaching format in the spring of 2020 forced more precarious and marginalized groups in academia to be further ostracized and isolated. They argue, however, that even with this forced move to a new videoconferencing tool, the chat function in Zoom provided a backchannel of sorts wherein precarious populations discovered a sense of belonging not found in the standard academic environment. The "networked togetherness" they describe (a transverse network, we would add) created a space that challenged the cis/het, white, rigid expectations of productivity in academia—demands that stayed firmly in place even during lockdowns at the height of the virus's impact. Heather J. Carmack, Heather M. Stassen, Tennley A. Vik, and Jocelyn M. DeGroot echo these concepts in "Stage Directions and Snarky Comments: Shadow Networks in Zoom Meetings Throughout the COVID-19 Pandemic," shifting their focus to the role of Zoom as a tool for administrators and faculty members charged with maintaining "business as usual" within a profoundly disruptive moment.[1] Through self-reflective scholarship, the authors explore their own pandemic experiences operating within these platforms and the moments within these networked interactions in which backchannels and shadow networks emerge as (transverse) paths within the same technological apparatuses that allowed videoconferencing to become the norm for academia. Della Ratta (2021) suggests that while videoconferencing platforms present themselves as tools of "collectivity and togetherness," one finds little more than a "hint toward shared spaces that do not exist, and the hinting begins from the very moment we call these spaces into being, from the moment we engage in the performance of their sociality." The chapters in this section, on the contrary, ask—in an educational setting at least—how that hinting might suggest not only a cue toward normative practices aimed at maximizing productivity but also minor modes of engagement that cut across and through these prescribed and prescriptive roles of the neoliberal (university) subject. Each chapter in this section is, while reflecting on "pandemic times," very much written *in medias res*, to the extent that many colleges and universities still rely heavily upon videoconferencing platforms such as Zoom to conduct meetings, perform student advising, and provide hybrid or fully online courses. This "new normal" for the neoliberal university is firmly in place. At the same time, these same platforms have not managed to obscure completely their precarious hold on networks of power and the degree to which transgressive openings and novel potentials for encounter still exist for students and faculty alike.

Note

1 That is not to say, however, that institutions did not "adapt" to the pandemic by revising expectations for grading (introducing pass/fail options), attendance, and even faculty scholarly productivity. The point that we make, rather, is that these accommodations operated as modulations within an ongoing control logic (Deleuze 1992) and one that was facilitated through digital tools and instruments such as Zoom that supported a logic of maximum performance.

References

Barad, Karen. 2007. *Meeting the Universe Halfway: Quantum Physics and the Entanglement of Matter and Meaning*. Durham: Duke University Press.

Deleuze, Gilles. 1992. "Postscript on the Societies of Control." *October* 59: 3–7.

Della Ratta, Donatella. 2021. "Teaching into the Void: Reflections on 'Blended' Learning and other Digital Amenities." *INC Longform*, January 6, 2021. https://networkcultures.org/longform/2021/01/06/teaching-into-the-void/.

Genosko, Gary. 2002. *Félix Guattari: An Aberrant Introduction*. New York: Continuum.

Gilles Deleuze, Gilles and Félix Guattari. 1987. *A Thousand Plateaus: Capitalism and Schizophrenia*. Minneapolis: University of Minnesota Press.

Guattari, Félix. 2015. "Transversality." In *Psychoanalysis and Transversality: Texts and Interviews 1955–1971*, translated by Ames Hodges, 102–20. Cambridge, MA: Semiotext(e).

Kuntsman, Adi and Esperanza Miyake. 2022. *Paradoxes of Digital Disengagement: In Search of the Opt-Out Button*. London: University of Westminster Press.

Sumner, Tyne Daile. 2024. "Zoom Face: Self-surveillance, Performance and Display." In *Performing Identity in the Era of COVID-19*, edited by Lauren O'Mahony, Rahul K. Gairola, Melissa Merchant, and Simon Order, 184–98. New York: Routledge.

Tomaselli, Keyan G. 2015. "Hacking Through Academentia: Autoethnography, Data and Social Change." *Educational Research for Social Change* 4, no. 2: 61–74.

7

The Zoom Machinic in Postdigital Learning Ecologies

An Exploration of Educators' Experiences via Three Case Studies

Kathryn Grushka, The University of Newcastle, Rachel Buchanan, The University of Newcastle, Michael Whittington, The University of Newcastle, and Rory Davis, The University of Newcastle

Introduction

> To "inhabit" technology, or to "welcome" it, would be nothing other than inhabiting and welcoming the finitude of sense . . . Rather, it is a matter of getting at the *sense* of "technology" as the *sense* of existence . . . The "reign of technology" disassembles and disorients the infinite feedback of a Sense. (Nancy 2003, 25–6)

The development of Zoom pedagogies is only now starting to be researched for its impact on learning for both teachers and students. Traditional measured learning outcomes tell one story but give little attention to the borderland between the virtual and material worlds of teachers and learners. To explore this liminal space, we draw on the analytic tool of "atmospheres" (Chandler 2011; Sumartojo and Pink 2019) and situate this analysis in an understanding that Zoom operates as a machinic assemblage. Learning on/with/through/via Zoom can be understood as taking place in an ecology where a multiplicity of human and nonhuman agents assembles together, and forces such as affective, social, technological, and natural relations give rise to our identities (Guattari 2008). A machinic assemblage contains energies such as sensation, perception, cognition, information, digitally produced images and sounds, interactive participation, and media—all of which shift away from an experiential world linked to nature.

The affect and impact of learning in a Zoom ecology is explored here through our collective autoethnographic analysis.

Sensing Zoom Learning Ecologies and Atmospheres

Postdigital learning ecologies acknowledge that the human and its machines have a relational connectivity with communicative potential, bringing greater levels of agency, social connectedness, and autonomy (Grushka et al. 2014, 2022). Applying Deleuze and Guattari (1987) to the Zoom environment, we see that the performative (or becoming) forces of a learning ecology involve an understanding of how our past and present experiential worlds of memories, technoworlds, and events collide and inform. Braidotti (2013) describes becoming as a continued folding and splitting of one's lived experience. In Zooming we become cyborg, a concept developed by the feminist philosopher Donna Haraway (1991), defined as "hybrid of machine and organism, a creature of social reality" (141). A student as cyborg is firmly located in the lived social reality of the learners' lifeworld. Both student and teacher draw on ways of knowing and being across multiple operational entities, multiliteracies, and techno-epistemological interactions and experiences. We ask, what does this digital experiential world of Zoom ecologies bring to our sensing selves and what learning insights does it generate when we inhabit the blended, new interdisciplinary borders between epistemologies, pedagogies, and learning spaces?

Applying the concept of machinic assemblage and machinic autopoiesis, we can come to understand the emergence of the concept of a Zoom learning ecology within a new materialist framework (Bennett 2010). Sociological ecologies have been defined within the biological human system. Here autopoiesis can be seen as active in maintaining stability in one's life through our interactions with other humans, families, and their communities. The traditional classroom is best understood in this context. Teachers via their classroom pedagogies focus on building a learner's capacity to interact, adapt, and learn. Its mode of sensemaking brings together perception, material being, cognition, and a collective, sociological knowing to support learning and the stabilization of our individual human sense of being and belonging for developmental well-being. A Zoom learning ecology, on the other hand, could be described as many and multiple "other" ecologies and ways of knowing that are co-opted into classrooms and homes as a messy entanglement of contemporary digital life.

Atmospheres are a way to understand peoples' changing, sensory, and affective experiences of spaces in the borderlands of learning from the virtual to the traditional classroom and the blended realities at the crossing over into and out of a Zoom learning ecology. The liminal space between materialities, cultures, and knowledge has been previously described in terms of the work of borders or borderlines in the dichotomizing of knowledge (Giudice and Giubilaro 2015). Deleuze and Guattari (1987) speak to how the borderlines are active within the assemblage, made up of bodies, actions, and passions, and the creative energies can either "de-territorize, cut the border and carry it away or re-territorize and stabilize it" (88). Zoom borderlines are potential spaces of productive force. Zoom ecologies carry new learning capabilities and learner agency. In the crossing into and out of Zoom atmospheres, the liminal spaces of blended e-learning are fluid and relational; they trouble past fixed dichotomies.

How, then, do we as actors with the agency afforded by Zoom ecologies do as Deleuze and Guattari (1987) argue and understand these spaces which are "continuously transforming [themselves] into a string of other multiplicities, according to [their] thresholds and doors" (249)?

The Inquiry

Post-structural, qualitative perspectives in research accept that knowledge can be generated from multiple perspectives. To reveal these complexities in the context of Zoom ecologies, we speak to our being "in" Zoom atmospheres, disclosing our affective being, questioning and probing individually and together. We inform one another via our experienced worlds, our unique places/spaces, and how we were participants within these spaces and the socio-techno imperatives of educational policies and pedagogies that surrounded and impacted our teacher praxis. We acknowledge that disclosing and probing lenses are brought into focus by affect:

> What you perceive or remember, what you try to figure out by reasoning, what you invent or wish to communicate, the actions you undertake, the things you learn and recall, the mental universe made up of objects, actions, and abstractions, therefore, all of these different processes can generate affective responses as they unfold. (Damasio 2021, 79)

The Zoom sites are the lived experiences of us as educator authors. We have drawn on an inductive, creative, process-oriented inquiry. The Gioia method, a reflexive, phased approach to qualitative analysis (Gioia et al. 2013), accommodates our individual and collective insights of the now commonplace Zoom machinic, postdigital learning ecologies. The Gioia method involves an initial phase of reflexive thematic analysis, which is followed by a collective meta-analysis of the first phase. Together we speak to multiple Zoom scenarios: being a learner in a virtual choir; taking a lead role with Zoom in Visual Art secondary school online learning contexts; and performing in the world of Zoom through the PhD supervisory process. The Gioia method allows the narrative voices of the authors to be positioned as "knowledge agents" (Gioia et al. 2013, 17) working within and across Zoom communities. As a conversational and reflective inquiry approach, it makes us cognizant of our own subjectivities as learners and teachers. It draws into focus how we had to disclose, question, probe, and rethink our own teacher identities as change and adaptation was a constant in all our experiences. It provided reflexive opportunities to talk about how our learners were experiencing the Zoom environment. Together we considered how we were all experiencing disruptive shifts in our subjectivities, and we all were aware that there was a need to reimagine how we approached e-learning opportunities.

The Gioia method inquiry has two phases:

Phase 1: Our voices as educators were inserted into the creation of Zoom ecology learning narratives. Sensed and lived experiences of teaching and learning were foregrounded, paying attention to emerging insights at interdisciplinary borders between epistemologies, pedagogies, and learning spaces. Background experiences and learning ecology insights were contextualized across the different online learning periods of 2020–2021. In the Gioia Method, this is presented as a 1st order analysis with the themes reflexively generated.

Phase 2: Here the analytical lens of knowing through atmospheres is applied. Knowing through atmospheres brings together the three sites with another analytical orientation. Together we examine diversity in a Zoom learning ecology. We considered whether normalized pedagogies had a place and creatively considered how we were able to break the limitations of past traditional pedagogies operating in Zoom and consider whether there were new, innovative opportunities for the learners and educators. In this contemporary educational learning space, the autoethnographic narratives consider Zoom possibilities and impossibilities.

Phase One: Narrative Voices

Kathryn: Becoming a Zoom Virtual Choir Singer

Becoming a virtual choir singer began in 2020 as Covid-19 forced the group online. As a member of an adult women's choir, the Song Sisters, it was going to take me some time to understand and be comfortable with experiencing singing life so differently. Knowing in a Zoom choir is different; somehow the flow, or our creative energies, was lost. I refer to flow as a state where I had spent years of joy creatively refining a unison sound made of multiple voices, being "with," being "a part of," and experiencing "in" the temporal vibrations of the voices of my chorister sisters. The qualities of Zoom learning ruptured how I had come to perceive and sense being "with" my choir. The richness of my known aesthetic and affective moments dulled. Singing in a choir is a collective and shared experience; we resonate and modulate together. We sing as one. We know our songs through experiencing one another's voices. Knowing about the limitations of Zoom singing was the first lesson. Learning new songs was always a slow and repetitive process but struggling through my soprano two part had always been punctuated with the collective joy of hearing our voices come together throughout a rehearsal. Zoom learning shifted these ways of knowing. I had to realign my perceptions, adjust my sensory and affective encounters, and learn to read my experience as broken, staged, and segmented. My embodied and situated familiar singing-self became unfamiliar. I now communicated with everyone via a digitally flickering object, the computer. I was a small square portrait, one of many on the screen.

Zoom design functionality continually repositioned us around the screen according to our unique voice intensities or our order of speaking. Sometimes I was visually present, other times absent. If you remained silent, you were not seen on the main screen without scrolling through participants. Our conductor, aware of this anomaly (a reality limitation), ensured that everyone introduced themselves before the start of a session. Latecomers remained silent. We learnt new songs via the screen in a very broken yet systematic way. In rehearsal, we never heard one another singing: just our own voice that floated in a solitary space and the sound of the conductor singing along through the audio speaker. My shared world of perception, sound, body, vibration, and proximal movement had been replaced with a digital screen of small-boxed portraits and the sole voice of the conductor in dialogue with the techno accompanist who used digitized recordings of the instrumental tracks, subsequently recorded over with the collective vocal. The choir could listen only to these recordings. The choir's social reality had now been disrupted by Zoom, the quality of our aesthetic encounters compromised. We knew our choir ecology as a physical "coming together" of a world of sounds, a collection of bodies, and the constant adjusting and modulating mass of voices in a particular place and space. Being in a Zoom choir ecology was a new virtual experience; the pedagogy and the pace remained similar, albeit modified, but image, time, and space were now controlled by the limitations of the digital device. Knowing was no longer a cacophony of voices and bodies that often overlapped, reorganized, or shifted in somewhat chaotic movement toward a refined unity of sound and emotion. It was not the same in the new virtual space; we were together but apart.

Rory and Michael: The Role of Zoom in Visual Art Secondary School Online Learning Contexts

Rory: The New South Wales State Department of Education (DoE) made the call in April 2020 for all staff to deliver one subject topic online. Children of essential workers were on-site but learned via Zoom. Staff struggled to navigate the platforms and to post consistent content. During that initial transition period, remote learning fragmented off into a range of delivery forms. Some scanned textbook chapters, others designed interactive modules with teacher supervision via Zoom. Staff were hosting video conferences across different platforms at inconsistent times, so to say it was overwhelming for students is an understatement. I encountered a new phenomenon that would become all so familiar when a chunk of the students "dropped off the radar." During the first lockdown, we had an engagement rate for thirteen- and fourteen-year-old students of around 45 percent. The DoE made three wholesale switches between remote and on-site learning between April 2020 and November 2021. From my perspective, I felt each switch had pedagogical refining potential, but our student engagement remained stubbornly low. There was a growing sentiment among us all that it was only temporary, not worthy of our full investment.

Student conversations conveyed frustration. Zoom lessons were spent nostalgically recounting practical artmaking lessons. In a visual arts studio learning environment, the materiality and participatory inquiry is palpable. We were missing the paint, the smell, the student movement, the flow, and rhythm of bodies performing self as constructed forms and drawing actions. There was an intense joy when we all were physically back together in the studio, coupled with a genuine fear, at least initially, of catching a deadly virus and taking it home to our families. It was the strangest dichotomy; I resented being forced back into the classroom, but I enjoyed being there. I wanted to be safe at home, but I felt deep sorrow at my lack of effectiveness. The most profound feeling from that period is being conflicted about how to adapt to the next shift. The possibilities of more blended learning loomed. Would it help or hinder? What will I keep or change?

Michael: In April 2020, the independent school I worked at was reluctant to allow any "wholesale" working from home. Instead, we built large online resources for students who were isolating at home that would run in tandem with the studio lessons. These would be put on all students' home learning pages administered by the school including the school's learning management system (LMS) Schoolbox, Google accounts, and YouTube pages. These processes were not a stretch as most visual arts teachers build large resources as part of their pedagogies as a regular practice. Our school was one of very few over those two years that did not shut down as a result of Covid-19. The school provided good Zoom pedagogical support. Students were expected to attend all their lessons live online via Zoom. It continued for two terms. There was a universal feeling and atmosphere of exhaustion. The excitement and novelty of Zoom had worn off in home learning. Students were all familiar with digital environments through informal and formal learning. They knew how to utilize the digital space to disrupt and subvert the process of learning. One example was students could very easily create fake anonymous profiles for Zoom lessons and students from other schools could Zoom in and create chaos. The platform had not been set up to manage these disruptions; however, the company quickly learned where its weaknesses lay through regular, self-perpetuating updates to minimize disruption and control student behaviors. It felt as if Zoom itself is its own autopoiesis. Over time the novelty began to wane and both students and staff felt Zoom fatigue settle in. Slowly, students were dropping out and not turning up to Zoom lessons. The excitement of Zoom play began to disappear. In its wake were disembodied, blank faces or a black tile with the student's name.

Rachel: Zoom and the PhD Supervisory Process

There is a photo on my computer sent to me and our PhD student, Gloria (not her real name), by a colleague. It is a screenshot from a supervision meeting via

Zoom. The picture shows an on-screen sample of Gloria's work beside the three thumbnails of us in the meeting. My colleague and Gloria are centered in their frames. In my frame, I'm off-screen, but my hand is visible pointing to one of the two plants on my desk. The plant that I am pointing to is large, thriving, and outgrowing its pot. The other is small and struggling. The brief email message that accompanied the picture was "Your thesis as a plant. ☺. Nice metaphor Rachel. You are making great progress Gloria." The reason for the message is that Gloria was doubting her capacity and progress. This meeting had started fraught, and she was doubting the reassurance we were giving her about the quality of her analysis. She was worked up and not hearing what we were saying to her. To convince Gloria that she was doing well, I had grabbed two plants from my office and placed them side by side on my desk. I had pointed to the smaller plant and said that many students at this point in their candidature were like that plant, but that her progress was akin to that of the larger, healthier plant. At that moment the tension broke, and she laughed, realizing that our encouragement was not being given as emotional support because she was upset, but that we were genuinely assessing her progress as excellent. That moment reset the emotion of the meeting. Gloria was able to relax and take in our words, and our discussion could move forward. My cosupervisor sent the message after the meeting to reiterate our encouragement and as a memento of an unplanned but effective teaching moment.

This supervision team has been operating via distance for the entirety of this candidature (as our student is based in another city), using Zoom and email predominantly, supplemented with the occasional phone call or text message. As an experienced supervisor, I find Zoom supervision harder than working with a face-to-face-based team. Trying to get to know someone in the context of a supervisory relationship is hard (but ultimately fulfilling) work. There is an emotional intensity to working with someone predominantly via screen. There is an intense dependence on fewer contextual cues to gauge how someone is feeling. There are awkward moments of silence as we try to determine who is going to speak or moments when we are talking over one another. In the moment described earlier, Gloria was not able to determine whether we were "just being nice" or whether we were being honest in our appraisal of her work. An in-person meeting would have had a quicker resolution of this situation, and I would not have had to resort to pantomime and metaphor to convince her of our intentions. While Zoom has allowed for the supervision of candidates for whom a move to my city is not practical, it requires a higher-intensity performance-based mode of teaching to overcome the problems that come with a lack of proximity.

Phase 2: Reflexive Analysis through the Frame of Atmospheres

Reflexive analysis allows for the theorization of Zoom as a machinic *in situ* and as a part of the digital experiential world and its learning ecologies. The analysis acknowledges the relationships of relationality and connectivity, both human and machine. Exploring being and becoming in Zoom ecologies reveals new theoretical insights into a technology that has transformed the use of videoconferencing.

Knowing in a Zoom atmosphere concerns itself with a focus on atmospheres as lived experiences. Here the autoethnographic lens required us to recall how we have come to know our own Zoom pedagogies: the ways we enter, explore, and leave these encounters; how others might experience the ways of knowing in Zoom. The qualities of the Zoom learning ecology engage perceptions, sensory encounters, and affective moments encapsulating the nature of computing in a Zoom world. In this Zoom world communication uses virtual design objects, visual modes such as video, film, media, animation, photography, and space-time/face-time conversations. This leads to us being able to describe and reflect across our own memories, imaginations, and the kinds of perceptual lenses we each brought to our Zoom encounters. Rory's reflections given here shows that Zoom is a collective endeavor. Students' attitudes to the encounter were shaped by that of the school, their teachers' access to technology, and the framing of being and becoming in Zoom situations:

Rory: Many conversations about assessments were prefaced by "when we get back" implying that the real work was to be done once on-site, in the "real" learning environment. I felt that the buy-in for my students was diminished, eroded further each time we ventured back to the classroom. I noticed a rebrand from "remote learning" to "learning from home." I assumed it was an effort to make it sound less isolating. I remember approaching the third shutdown feeling that we were teaching "through" technology, rather than with it. In my context the prevailing attitude of students was that Zoom (and Canvas, our LMS) were gatekeepers that compounded issues for our community regarding equity and access to technology. My evolving key takeaway was that our school has a wide range of cultural and social backgrounds, and each round of "learning from home" highlighted the yawning gap in student access to technologies at home.

Michael: I had to swiftly adapt the pedagogy to digital portraiture as most students' home learning environments were somewhat limited. For example, there were no art materials in most students' homes, and some were lucky to even have a pencil.

Rachel: I use my own emotions to gauge how a meeting went. After a Zoom PhD supervision meeting, I am often exhausted, more tired than I would have been had that meeting been face-to-face. When on Zoom I repeat myself more. I squint and strain my eyes trying to keep the visuals on the screen in focus. I overbook myself with back-to-back meetings. I screen share and talk through the annotations I have made on students' work, hoping my comments are helpful and productive for the student. Checking in with the student is harder without the perceptual clues I would get from being in the room with them. The camera and the screen can lead me into a tendency to monologue. The give and take of a face-to-face conversation can be inhibited by Zoom. I find myself smiling and nodding emphatically to ensure that the student is aware of my encouragement and ongoing support. I perform a little more to compensate for the lack of proximity.

Rachel's lived experiences speak to the fatigue that can come with Zoom teaching. She knows her pedagogical repertoire and response to the Zoom ecology and is able to identify these qualities, heightened affective moments, and performative compensations that come with Zoom teaching. For Kathryn knowing in a Zoom atmosphere brought a sense of loss. She was keenly aware of what was different and lacking in comparison to the sensory richness of the in-person choral experience. Michael identified that in a Zoom atmosphere the needs of his students were different, and he had to adapt his teaching.

Knowing about the Zoom Atmosphere required us to seek out and explore the ways we approach defining Zoom atmospheres retrospectively (Sumartojo and Pink 2019). We acknowledge that capturing these retrospective feelings and moods has moments of impossibility. This perspective can help us communicate the experiences of being in a Zoom environment, its limitations, and how the participants can work with these limitations to design new processes and pedagogies. We drew on our own past experiences and what policy and practice told us about the benefits and limitations of this environment. We imagined what Zoom atmospheres might be (Sumartojo and Pink 2019). From an analytical perspective, we focused on reflexivity to interrogate the qualities of these learning encounters and their aesthetic and embodied significance. Together we actively talked about what teaching was like in Zoom. We disclosed our failures, and we questioned and examined our judgments and encounters as we sought to eliminate preconceived conclusions in ways that supported open engagement with our shared realities.

Michael: Students in the home learning Zoom atmospheres were suffering from underdeveloped 2D portraits. I needed to keep their engagement active. YouTube tutorials were limited. Students had to rely on teacher feedback once their drawings were completed. It was difficult to intervene and give formative feedback as the work was not seen in process. This means that the teachers' artistic intuiting skills were limited behind the screen. Who could really see what was going on with their hands? Yet at the same time students who by reputation would normally spend much of their personal and social time in these online and digital spaces thrived. One student had his study room decked out with selfie lights, a wall full of Funko Pop vinyls, and other movie, gaming, and tv memorabilia. The student also had a headset that allowed his voice to be clean and well-articulated. He would often ask me for permission to help me with my demonstrations by asking "Sir, can you make me co-host?" This student was one of the quiet ones in the studio. In this online space he was displaying confidence. His digital pieces were outstanding, but his traditional drawings needed reworking. What was it about the affordances of this space that allowed him to confidently create work—and keep his morale up? The students in the Zoom atmosphere would often ask me at the end of the lesson before I signed off to make them co-host so they could spend their recess or lunch time socializing via Zoom. I was cautious about this but allowed it as this was their only real social time, and I trusted these students.

Rachel: Reflection on the experience of Zoom supervision and comparing it to face-to-face encounters with my students allows me to articulate the layer of difficulties that supervision via Zoom presents. While Zoom allows for flexibility, and the capacity to work with students who are not based on campus, Zoom meetings are slightly more fraught, more of an exaggerated performance and a little less comfortable. I am less able to gauge the emotion of the student. I use repetition and reassurance to reinforce what is being said.

Rory: Absence from the classroom necessitated the final practical task to be completed using equipment readily available to home. Visual diaries had been delivered to students early in the lockdown. The portfolio of work was to be about an exploration of home and garden. It was an attempt to connect them to their real-life experiences using their natural surroundings as stimulus. A few events forced pedagogical adaptations, and their impact was felt across the wider school. Firstly, my student online numbers were now very low, some had "dropped off the radar" completely. Secondly, I had shifted to "mute" studio learning in which we watched YouTube then muted ourselves in our own studios to draw off-line. Thirdly, mental health issues were on the increase across the school. It was no surprise when the school stopped online home learning.

These reflections show how as educators we were aware of limitations in terms of material resources, access to technologies, plus the teachers' knowledge of pedagogical strategies. Michael's example shows that for some the digital enhanced rather than diminished their learning experiences.

Knowing through Zoom Atmospheres brings together the things and processes that require examination within a Zoom ecology. It connects the theoretical lens of the machinic assemblage of Zoom. It helps us describe how entering the space-time of Zoom experiences revealed new conceptual insights:

Rory: The mood at first was one of curiosity. I adapted a tutorial I had found online. After a quick check in I would hold up to the camera the drawing equipment I had selected, and we would compare our favorite artmaking tools we had available at home before starting the tutorial. We left our cameras on whilst I streamed the video through Zoom. It felt immediately familiar, each of us busily working in our visual diaries and occasionally showing others our works in progress, we problem-solved together. Familiar self-deprecating jokes were offered, followed by sincere compliments from peers and an overall feeling of reciprocated kindness. It was a morale boost. By the end of the calls, we all had a level of elation and excitement. My students were soon anticipating their next artmaking session.

Michael: Zoom learning is good for students who can adapt and grow with its developing interface. I certainly would have been lost without it. I do not think that the students would have done so well without the learning environments we had, given the circumstances.

Rachel: Experiential knowledge of the affordances and the limitations of Zoom has led to a shift in my supervisory pedagogy. Working in the shared online space of the Zoom meeting, the screen is a partition, and the camera creates focal points that differ from those in a shared space. Screen sharing leads to a different form of marking up students' work—annotations using Word Review software, rather than the idiosyncratic writing on the page that I would otherwise use. Without the freedom of pen to paper, my feedback is more considered and more economical. This is a pedagogical shift in the way that I give feedback. Screen sharing also means that the student sees this feedback while they themselves are on screen, creating moments where they may need to modulate their emotion, rather than being able to reflect on such feedback in private. These shifts reveal the impact of the machinic Zoom ecology.

The affective shifts that come with Zoom ecologies can be sensed in the previous reflections. Moments of peace can come through the shared Zoom experience, but this also brought a longing for a return to the sharing of the physical classroom. The collectively felt impact of being together but separate, coupled with a technologically mediated mode of interaction, shapes the Zoom atmosphere.

Discussion: Digital Zoom Ecologies and Knowing through Atmospheres

Zoom has emerged as a marker of the contemporary *zeitgeist*. The theoretical work of this chapter has relevance as it allows a rethinking of the liminal spaces—atmospheres, the borderlands—between educator and students on the other side of the screen. We considered how emotions generated from digital perceptual events are described. Repetition, boredom, and fatigue, both visual and physical, do impact on cognition and learning (Damasio 2021). Postdigital learning platforms reveal many insights about human perception and memory in foregrounding the multimodal cognitive benefits of techno-learning ecologies. Zoom is presented as a significant tool in online learning, particularly in higher education. The benefits purported do so without recognition of what is lost or its role in the interplay with the materiality of being and learning. Post Covid-19, Zoom realigns our temporalities. Zoom recordings of lessons can be accessed in isolation after the event. The poignancy of face-to-face teaching, where reflexive moments and questions are explained and emphasized, can never be repeated. In contrast, the recorded Zoom lesson—rehearsed, revisited, and revised in the name of assessment—aligns strongly with the performative logic of instrumental, outcomes-based pedagogies.

What does this digital experiential world of Zoom ecologies bring to our sensing world, and what is gleaned as insights into the new, interdisciplinary borders between epistemologies, pedagogies, and learning spaces? What might be adapted for individuals who inhabit Zoom spaces? To adapt to the Zoom space is to get used to the changed visual environment of the screen. It involves finding ways to maneuver digital spaces within the constraints of a lack of physical proximity. For some it involves getting used to being recorded—a shift from teaching being an ephemeral act to being available for posterity. In inhabiting Zoom spaces teaching becomes more entwined with technology; the materiality of papers and pens is replaced with digitally marked-up documents. In the arts classroom, specialized equipment for artmaking is replaced by what students have available in their homes. In a postdigital world, the learning spaces of analog and digital are no longer two different things (Cramer and Jandrić 2021). In the collapse of these two worlds, the challenges of educational politics mandated a switching between digital and analog environments. It was a dichotomous experience for Michael and Rory. In this paradigm, it is very easy to see the benefits of a blended learning environment where both can be encompassed. This research, however, demonstrates the pedagogical realities that confront the visual arts educator. The economic benefits that come with digital learning (versus resource-intensive studio learning pedagogies) compromise the pedagogies of visual art teaching. While visual arts educators offer both material and immaterial digital expressive learning mediums, from animation to painting and drawing to video, learning impossibilities persist. The studio participatory learning space is compromised when the pendulum shifts toward an emphasis on online learning.

Relationships can be developed and maintained within Zoom learning technology, albeit with a physical connectivity replaced by technological connectivity. That the show (or the song, or the lesson) must go on attests to the adaptability and resilience of teachers and learners in the Zoom ecology. These studies show that students and teachers can have capacity for growth using Zoom—they find ways of making things work or coping with unexpected and unfamiliar experiences. Dependence on Zoom means a dependence on a raft of technologies (electricity, software, hardware) and processes beyond the home and classroom. When the internet is out, the experience is inaccessible. Zoom atmospheres foreground the significance of both affective interdependence and the technological. Choral practice can be refigured so that singers can be together while apart. Classroom studio pedagogy can evolve into the home and garden becoming the studio. Yet our experiences suggest that Zoom may have blunted some very particular ways of knowing and associated learning modalities.

What is lost or gained in the new liminal space of meta-production must be reconciled to be successfully inhabited. The complex relationships, connectivity, and interdependence for teachers and students working in a Zoom learning ecology to build individual growth require adaptation and resilience. Can this be maintained for our diversity of learners? Will we blunt the perceptual and affective experience necessary for quality cognitive encounters? Atmospheres allowed the inquiry to consider ambiances, feelings, and the quality or mood in the Zoom learning moments as well as other sensory experiences such as vision, sounds, movement, and their qualities such as speed and tone. Learning draws on our increasing awareness of embodied knowing, human perceptual/ sensory limits, and the significance of feelings in the conscious mind and in the management of life (Damasio 2021).

Zoom Contemporary Educational Learning: Disrupted, Flattened, and Controlled

Contemporary learning systems in education form collective machinic assemblages in society and work as an element of both control and surveillance. Control is explored from the perspectives of disrupted creativity and flow. Flow in learning is where one's senses are heightened, time is lost, and learning is intense and focused (Csikszentmihalyi 2013). More mundane goal-centered learning, such as Zoom pedagogies, do not have these qualities. Ferng (2020) speaks to the nature of the architecture in post-Zoom ecologies as a flattening/smoothing and diminishing of the aesthetic, sensory, and material experience. In postdigital societies, learning technologies can operate as a means of monitoring and as tools of persuasion and control (Buchanan 2020; Buchanan and McPherson 2019). In many classrooms, students are monitored, their results, interactions, and behaviors recorded, tracked, and analyzed. Such practices "inscribe children within an ever-intensifying network of visibility, surveillance, and normalization, in which their behaviors and bodies are continually judged and compared with others" (Lupton and Williamson 2017, 7). In the Zoom environment students can feel this surveillance, and practices such as not turning on their cameras can be an act of resistance.

Zoom screen architecture allows access to diverse, sensory-rich digital information from individual desktops, making Zoom an effective medium for interaction. However, it comes with considerable structural and aesthetic limitations. Ferng (2020) refers to Bertolt Brecht's "alienation effect" (208). Zoom is an artificial environment where the phenomenological complexity of our vibrant world is received as flat, flickering digital images. The atmosphere of alienation, stress, and lack of rich learning is echoed in our accounts of work through Zoom. The screen produced a minimal voice, while opportunities to explore visual effects abounded. "Staring at the screen for long periods of time only seems to generate greater amounts of ennui, draining our creativity and productivity" (Ferng 2020, 208). Participants and students spoke to this kind of exhaustion and a longing to return to the "real" classroom. The human/machine interface and its atmosphere of Zoom is not an equivalence of our biotic perception as we engage with real-world materiality. It can, however, still work as a technology of social connection.

The enticing world of visual perception and the richness of place, space, shape, and color found in the dynamic real world are diminished by the digital Zoom atmosphere. In Zoom environments voices are muted, flattened, genericized, unable to be performed. "Talking heads" stream from offices, homes, and classrooms, or just as often: in front of fake digital backgrounds that disturb even the Zoom reality. The dominance of text-based sign systems of sound remains, and any "reading" of the face, gesture, and body is muted or flattened within Zoom. DeLanda (2022) describes this perceptual phenomenon as graded consciousness, whether in real time or recorded digital time; both contain temporal anomalies and visual variants. Zoom is a biotic/abiotic intersection, a narrowly streamed reality. This flattened avatar Zoom learning environment becomes one of many digital encounters that are increasingly producing learner weariness.

The machinic assemblages of Zoom worlds generate the disequilibrium of the virtual universe. Applying a Deleuze and Guattari (1987) lens we see everyone with the ability to affect and be affected and indeed adapt within this Zoom ecology. In Zoom learning, meaning making emerges in the assemblages of nonhuman and human actors, a gathering together of bodies, materials, histories, and enunciations. While Zoom has been a blessing for many in isolation, whether pandemic-induced or a result of distance learning, it has also reasserted teacher-centered learning and control, reflected by both Rory and Michael as the "loss" of studio learning.

Zoom was designed for corporate rather than educational applications. While many teachers have made productive use of Zoom, the affordances of the technology constrain educators' pedagogical repertoires. The impossibilities of collective, large group learning when students are on Zoom together are problematic in the same way that a choir is unable to collaboratively learn. Many students simply mute themselves and simply disappear (Bruzzone 2021). There also remains a hierarchical value placed on such signifying behaviors.

Conclusion

This chapter sought to utilize the experiences of the authors across three learning sites. The aim was to explore the possibilities of new epistemological insights surrounding Zoom ecologies. An atmosphere analytic of the three cases suggests that although Zoom offers pedagogical possibility, other opportunities are lost. The findings suggest that we need to be wary of a technology of communication that begins to limit the rich possibilities of the materiality of our bodily "real-world" learning experiences—where spatial moments carry collections of bodies and the chaos and magic of performative figures that sense, smell, speak, and sing. Human touch disappears as we are all reduced to the low constant vibrating neutral whir of the machine; the smell of plastic and metal; the touch of fingers sliding over metal and glass; heat and the digital flicker of a torso telling stories.

The digitalized and flattened reality of the Zoom classroom changed the dynamics of the choir, the visual arts classroom, and the PhD supervisory experience. Neuroscience helps us understand the significance of a feeling, remembering, rational mind in the postdigital world. The application of postdigital theory, atmospheric methods, and neuroscience insights into the significance of perception and feeling combined with intuitive and reflexive thinking has allowed open consideration of the pedagogical limitations that were encountered. This chapter speaks to the generative frictions in Zoom ecologies of learning. It asks for a reconsidering of technophilic or technophobic positions that characterize much of educational technology research. Contemporary learning systems in education present these denaturalized learning ecologies defined by intermediality, temporality, and social cognition. Zoom has many different sensory modalities of interaction better understood within the "ongoingness" of collective machinic assemblages that work as an element of a control society. Rather, it is a matter of getting at the *sense* of "technology" as the *sense* of existence. The "reign of technology" disassembles and disorients the infinite feedback of a Sense (Nancy 2003, 25–6).

References

Bennett, Jane. 2010. *Vibrant Matter*. Durham: Duke University Press.

Braidotti, Rosi. 2013. *The Posthuman*. Cambridge: Polity Press.

Bruzzone, Silvia. 2021. "A Posthumanist Research Agenda on Sustainable and Responsible Management Education after the Pandemic". *Journal of Global Responsibility* 13, no. 1: 56–71. https://doi.org/10.1108/JGR-05-2021-0045.

Buchanan, Rachel. 2020. "Through Growth to Achievement: Examining Edtech as a Solution to Australia's Declining Educational Achievement." *Policy Futures in Education* 18, no. 8 (March). https://doi.org/10.1177/1478210320910293.

Buchanan, Rachel and Amy McPherson. 2019. "Teachers and Learners in a Time of Big Data." *Journal of Philosophy in Schools* 6, no. 1: 26–43. https://doi.org/10.21913/JPS.v6i1.1566.

Chandler, Timothy. 2011. "Reading Atmospheres: The Ecocritical Potential of Gernot Böhme's Aesthetic Theory of Nature." *Interdisciplinary Studies in Literature and Environment* 18, no. 3: 553–68. https://doi.org/10.1093/isle/isr079.

Cramer, Florian and Petar Jandrić. 2021. "Postdigital: A Term That Sucks but Is Useful." *Postdigital Science and Education* 3, no. 3: 966–89. https://doi.org/10.1007/s42438-021-00225-9.

Csikszentmihalyi, Mihaly. 2013. *Flow: The Psychology of Happiness*. London: Random House.

Damasio, Antonio. 2021. *Feeling and Knowing: Making Minds Conscious*. New York: Pantheon Books.

DeLanda, Manuel. 2022. *Materialist Phenomenology: A Philosophy of Perception*. London: Bloomsbury Academic.

Deleuze, Gilles and Felix Guattari. 1987. *Thousand Plateaus: Capitalism and Schizophrenia*. London: University of Minnesota Press.

Ferng, Jennifer. 2020. "Post-Zoom: Screen Environments and the Human/Machine Interface." *Architectural Theory Review* 24, no. 2: 207–10. https://doi.org/10.1080/13264826.2020.1809769.

Gioia, Dennis A., Kevin G. Corley, and Aimee L. Hamilton. 2013. "Seeking Qualitative Rigor in Inductive Research: Notes on the Gioia Methodology." *Organizational Research Methods* 16, no. 1: 15–31. https://doi.org/10.1177/1094428112452151.

Giudice, Cristina, and Chiara Giubilaro. 2015. "Re-Imagining the Border: Border Art as a Space of Critical Imagination and Creative Resistance." *Geopolitics* 20, no. 1: 79–94. https://doi.org/10.1080/14650045.2014.896791.

Grushka, Kathryn, Rachel Buchanan, Michael Whittington, and Rory Davis. 2022. "Postdigital Possibilities and Impossibilities Behind the Screen: Visual Arts Educators in Conversation about Online Learning and Real-World Experiences:

Visual Pedagogies and Blended Learning." *Video Journal of Education and Pedagogy* 7, no. 1: 1–23. https://doi.org/10.1163/23644583-bja10027.

Grushka, Kathryn, Debra Donnelly, and Neil Clement. 2014. "Digital Culture and Neuroscience: A Conversation with Learning and Curriculum." *Digital Culture & Education* 6, no. 4: 358–73.

Guattari, Felix. 2008. *The Three Ecologies*. London: Continuum.

Haraway, Donna. 1991. *Simians, Cyborgs, and Women: The Reinvention of Nature*. New York: Routledge.

Lupton, Deborah, and Ben Williamson. 2017. "The Datafied Child: The Dataveillance of Children and Implications for their Rights". *New Media & Society* 19, no. 5: 780–94. https://doi.org/10.1177/1461444816686328.

Nancy, Jean-Luc. 2003. *A Finite Thinking*. Stanford: Stanford University Press.

Sumartojo, Shanti and Sarah Pink. 2019. *Atmospheres and the Experiential World: Theory and Methods*. New York: Routledge.

8

"Networked Togetherness, I Guess ¯_(ツ)_/¯"

Subverting the Academic Zoom Chat through the Subcultural Collective

Alexis-Carlota Cochrane and Theresa N. Kenney

Chat

Alexis-Carlota to Theresa (Direct Message)
PLS LET'S ZOOM. 💻 💗 Need friend screen time soon . . .
also need to meet deadlines tho 😂 😂 😂
1:54 pm

> **Theresa** to Alexis-Carlota (Direct Message)
> OMG ready for it. 🙌 SO HYPED
> 1:55 pm

In the case of the university, March 2020 brought on a swift shift to the virtual. The coronavirus pandemic necessitated digital pedagogical methods on, at least for academics, new videoconferencing platforms. With its virtual meeting rooms, built-in collaboration tools, recording capabilities, and chat function, Zoom became a crucial digital platform for learning and meeting online. These social and digital affordances (Graves 2007; boyd 2011) promised a means to replicate in-person university experiences. Remote undergraduate course delivery could still offer synchronous lectures, small collaborative group work, one-on-one meetings, and pop quizzes, even if mediated by screens. Scrolling chat boxes recalled the whispers of the lecture hall, and raised platform-integrated hands emojis (✋) mimicked the gestures of eager students in the in-person classroom space. Lecturers could continue to use familiar slideshow presentations through screen sharing all while monitoring attendance through participant login lists.

Meanwhile, in the realms of graduate learning and training, Zoom functioned to connect graduate students with their supervisors, peers, and collaborators. The platform sustained graduate research progress and permitted virtual dissertation defenses in addition to combating the already overwhelming isolation of master's and doctoral studies. "Zoom University" (Aratani 2020) offered a semblance of in-person, pre-coronavirus teaching and learning despite spotty webcams, thumbs up emojis (👍), "can you see my screen?" questions, virtual backgrounds, "your mic is off!" interruptions, and technical difficulties.

Practically overnight, many of us in academia were propelled into what has come to be called the "new normal": virtual work in back-to-back Zoom meetings, sitting for hours in the distinct glow of blue light screens, and increasing isolation from the outside world. For many of us who are already positioned as precarious, low-income workers of the university, as we will describe in the pages that follow, our days consisted of waking up and taking a few steps to our small workspaces—desks pushed into the corners of rooms, kitchen tables, and couches. We taught virtually as disproportionate death loomed, which, as Saidiya Hartman reflects, involved "pretending to prepare for class, complying begrudgingly with the university's demand to proceed with business as usual" (Hartman 2020). We suggest that this similitude is worth troubling since a time of crisis should not be so normal. We contest it and the many ways such normality upholds inequitable expectations of performativity, professionalism, and precarity as ordinary. Crisis surely warrants conditions, methods, and attitudes that differ from the norm, perhaps especially in academia. And yet, as Wendy Hui Kyong Chun reminds us, "neoliberalism thrives on crises: it makes crises ordinary" (Chun 2017, 3). There is nothing "new" about the cultures of rigor and production, especially in academia, and still, the shift online continues to amplify production-focused economies.

Striving to provide a virtual version of in-person delivery caused the institution of the university to falter to its ordinary pedagogical methods and expectations, which merely transferred already unsustainable academic working and learning conditions onto inherently inequitable digital platforms during a time where "everyday existence [was especially] on the verge of catastrophe and expected disaster" (Hartman 2020). At the institutional level, the university relied on "the intertwining of crisis, capitalism, and platformization" to optimize overburdened operations and sustain pre-pandemic levels of revenue (Grandinetti 2022). More than ever, technocapitalism (Suarez-Villa 2009) enabled the same old pedagogical practices to stay complicit with neoliberal "indicators of success" for teachers, researchers, and students, including enrolment figures, grant-driven "revenue streams generated by university research and development[,] and the ability to attract higher-paying students" (Luka et al. 2015, 177). On the individual level, the university became easily accessible in the (dis)comfort of our own homes. Quarantine measures did not stop demands to keep up normal academic trajectories (see Kenney et al. 2023). Arguably, the shift to online environments more so heightened the performance-based surveillance of academic norms through the continued adherence to the methodological and professional rigidity of the ivory tower. As queer graduate students of color who are still primarily using Zoom for our training and research, we are all too familiar with the "perpetuity of the old (normal), structural anomalies [of the university that] continue to haunt" us online (Pathak 2022, 64). And on the Zoom platform, precarity, performance, and pressure abound.

Still, Zoom can enable all sorts of subversion. In this chapter, we consider the ways in which the Zoomscape (Breen 2021) might also act as a vehicle for queering the expectations (Hunt and Holmes 2015, 156) of normative academia: its excessive rigor, demanding professional training, and cultures that maintain the institutional status quo. Like Stefanie Duguay et al. (2022), we engage with how "Zoom holds promise for intimate uses"(6). Through several conversational epigraphs reminiscent of direct message Zoom chats between the authors, we model "a method, process, and product" grounded in our own experiences of networking together digitally (Batac 2022, 68). These recurring epigraphs are our attempt to exemplify how autoethnography can both be data *and* inform analytics (Zeffiro 2021, 453) in digital platform studies. At the same time, the epigraphs intend to emphasize the visuality of the Zoom chat (the platform itself, its emojis, and its potential for colloquial conversation) all while demonstrating how collegiality can be garnered in such complex online spaces. Recalling the uprise in chat applications, videoconferencing platforms, as well as text-based social networking services(SNSs), and the scholarship on these online platforms (boyd 2011; Jenkins 2016; Hill 2018), we attend to the ways in which subcultural humor, memes, and emojis can queerly subvert the expected uses of Zoom for both academics and the neoliberal university itself.

We recognize that the "pressure always to assimilate" within academic spaces "manifests differently and more severely" on those who are "[B]lack and woman and queer and [disabled] and immigrant and [. . .] on bodies that cannot be assimilated" (Antwi 2018, 310). These are those who were already marginalized in academe pre-pandemic and unevenly impacted (Hartman 2020; Jiwani

2020) by pandemic times. Therefore, following scholars like Stefano Harney and Fred Moten (2013), who have written on the collectivity of what they call "the Undercommons," we consider fugitivity in university online spaces in order to recognize the togetherness of those—the colonized, racialized, queer, immigrant, neurodivergent, and otherwise precarious—who do not perform the rigidity or sameness of the institution. We propose the concept of *networked togetherness* to name digital processes, practices, and pedagogies that foster and are fostered by a network of those—the colonized, racialized, queer, trans, immigrant, neurodivergent, and otherwise precarious—who engage collectively and subversively to nurture belonging, enact survival skills, and craft community in-and-through subcultural references and commitments to care. To conclude, we assert that subcultural humor is a critical intellectual tool for using the subcultural collectively and togetherness itself acts as a form of resistance to the strictures of academia, especially at times of crisis and online. We contend that, in spite of its complicity with neoliberal responses to crisis, the Zoomscape does "enable new forms of pedagogy, resistance and, most fundamentally, survival" (Hill 2018, 292) for those networking together through subcultural and resistant means like humor. In turn, we explore how the Zoom chat can destabilize dominant and normalizing cultures of academic rigidity insofar as it can open avenues to challenge capitalist productivity and so-called professionalism, despite the many ways online platforms demand otherwise.

Chat

Alexis-Carlota to Theresa (Direct Message)
me: reading your kind comments
ahead of our Zoom meeting tomorrow
11:01 am

> **Theresa** to Alexis-Carlota (Direct Message)
> WAIT!? we have a meeting tomorrow? WHAT
> 11:02 am

Alexis-Carlota to Theresa (Direct Message)
😵 💀 😵
11:02 am

> **Theresa** to Alexis-Carlota (Direct Message)
> This is fine
> 11:05 am

While our interests here are to analyze how digital mediations of the Zoom screen perpetuate both complicit and subversive practices, we must acknowledge how technosocial environments—user interfaces, platforms, and programs formed out of code, algorithms, and software—are shaped by (and shape) systems of power. Technologies are built to reproduce white, colonial, capitalist, anglophone, cisgender, straight, able-bodied, and neurotypical logics (Risam 2018). On one hand, digital technologies and their infrastructures require mineral extraction (Chen 2012; Yusoff 2018)—copper for wiring, lithium for batteries, cobalt for circuit—in ways that necessitate Global South labor predominantly done by people of color—in Brazil, India, Guinea, Indonesia, Jamaica, and so on. Such methods of geologic and labor extraction continue colonial and capitalist models of exploitation of both the Earth and people of color. On the other hand, hidden behind the falsehood of technological neutrality are the embedded discriminatory norms of whiteness, masculinity, heterosexuality, neurotypicality, and so on. These technologies perpetuate social inequalities, reinforcing various marginalized realities in digital spaces. To recall Safiya Umoja Noble (2018), this replication of social and cultural power dynamics "is not just a glitch in the system but, rather, is fundamental to the operating system of the web" (10). Therefore, technologies are always already implicated in the transnational technocapitalist circulation of people, products, and ideas, linking digital culture with ongoing histories of white supremacy, extraction, displacement, and stolen land.

Zoom is certainly no exception. Zoom Video Communications, Inc has material ties to capital; it is a proprietary corporation with a transnational presence as a productive and innovative digital landscape for remote working and learning. And yet, describing digital platforms as "environments" and "landscapes" neglects the very real ways these technologies are situated on and indebted to land. In the case of Zoom, it is the land of the Ohlone that sustains the social and technical infrastructures of Silicon Valley that make its digital meeting rooms possible. Moreover, it is settler colonial institutions such as universities that subscribe, quite literally, to such platforms, in turn, replicating and reinforcing other technologies of power. Zoom is then irrevocably bound up in the long and various histories of Indigenous genocide as well as in what Ruth Wilson Gilmore (2007) describes as California's long history of "interlacing the military complex with consumer and producer goods manufacturing, agriculture, resource-extraction industries, and high levels of consumption" (37). And the university's attempt to replicate academic working, living, and learning conditions online only amplifies how digital and nondigital landscapes are dependent on norms that are inextricable from larger systems of violence, including white supremacy, ableism, and settler colonialism.

If power structures and frames technology, then it also mediates how users experience digital platforms. Zoom, for instance, builds on the functionality of platforms and applications of Web 2.0, offering opportunities to interact virtually across its various elements like the chat function. In comparison to the "static" consumption of Web 1.0, Web 2.0 brought on a participatory internet culture that was not present in the previous iterations of the World Wide Web (DiNucci 1999, 32–5). Users with advanced digital literacy can easily access digital affordances on Zoom such as chat, emojis, reactions, live transcripts, and so on due to their skills, knowledge, and resources on earlier iterations of social networking and microblogging websites MySpace, Friendster, Pizco, and Tumblr. These are users who have material experience with computers, the internet, and virtual interfaces, which has long been shaped by "the cost— both financial and cultural—of acquiring Net literacy" (Nakamura 2002, 108). Many other users "especially many of those economically deprived, socially marginalized, historically disempowered" (Jenkins 2016, 11) cannot so easily approach and access these platforms. While we invoke access here in terms of material and cultural capital to mark how racialized, gendered, and queer users may experience Zoom differently, we also conceive of access through critical disability frameworks that, in the case of the virtual, contend with what Ashley Shew (2020) calls as "technoableism" (41). In other words, virtual spaces are generative platforms to replicate disparities, reinforce rigidity, and reinscribe violence as related to race, sex, gender, and ability. It is unsurprising then that the university sufficiently transferred to such online platforms to maintain its neoliberal aims.

It is less surprising, however, that two queer, neurodivergent femmes of color approach technology through care, access, anti-racism, and queerness. We are compelled by practices that "resist the gravitational pull towards heteronormativity and binary thinking in the ways technologies are imagined, narrativized, redesigned, and used" (Shaw and Sender 2016, 1). As two people who grew up on the internet, we have come to know (see Zuroski 2020) how to play with various platforms affordances and constraints. The introduction of SNSs in the late 1990s shifted user roles from passive bystanders to active participants. The internet became interactive with customizable user profiles, guestbooks, personal blogs, and chatrooms. While all SNSs have digital affordances designed to orient their users to engage with others—follow friends, post updates, create hashtags, share pictures, favorite posts, and more—these digital social networks remain, to some extent, because users generate their own engagements with one another, facilitating all sorts of community building through connection. As users reinterpret, adapt, and reinvent (Eglash 2004) digital platforms, and do so in community with others, they form counterpublics or backchannels that challenge mainstream discourse and expected participation online.

Hill (2018) extends Michael Warner's (2002) writings on counterpublics in order to consider the continual innovation of communication technologies like SNSs and an increased access to the internet. With a particular focus on Black Twitter as a digital counterpublic, Hill (2018) suggests that online platforms have pushed counterpublics to relocate online, shifting "notions of community and publicity" as well as collective engagements in political activism, community, and resistance (289). Considering the ways in which platforms surveille Black users, he looks to how these users have also mitigated or repurposed engagement through counterpublic practices of critique, awareness, and care. For Hill (2018), these digital counterpublic practices make Black Twitter "a radical site of feminist and queer politics and pedagogy" (297). We undertake a similar approach to Hill in our analysis of Zoom, exploring how this digital platform can facilitate an alternative to academia's current structures, enabling "new and transgressive forms of organizing, pedagogy and, ultimately, resistance" (Hill 2018, 297).

Of these forms, we attend to the network (Chun 2017) to recognize the overlaps shared between networked technologies (boyd 2011), digital networked counterpublics (Renninger 2015), and networked togetherness. The network indexes multiple technologies, communities, and practices. Yet, unlike other forms, the network is as much about the process of its formation as it is about the form itself. Put another way, the process of networking—constellating, coming together, assembling, arranging, engaging, and so on—determines the functionality (and fugitivity) of the network. "Networked togetherness" echoes long-standing practices shared within and between Black, Indigenous, and other communities of color offline. These practices bring together the fugitives—those who are thinking, surviving, stealing, hiding, and laughing tactically. They are coalitional and subversive practices based on a shared recognition "that it's fucked up for you, in the same way that we've already recognized that it's fucked up for us" (Harney and Moten 2013, 140). At the same time, networked togetherness is inherently about care in part because care is as strategic and subversive as survival and resistance. Radical care is "a set of vital but underappreciated strategies for enduring precarious worlds," which provide otherwise world-making "even if it cannot completely disengage from structural inequalities and normative assumptions" (Hobart and Knesse 2020, 2–3). Care demands an attempt to divest from individuality and isolation. It is a politics of relation that de-emphasizes "competition for institutional resources or recognition" and instead embraces modes of "togetherness" (Finley 2020, 367–8). Networked togetherness, we argue, is one of these modes.

Moreover, the resistance and care-fullness of a networked togetherness is arguably queer. On one hand, it is a mode of togetherness forged between those who share a marginal position to power. If queer refers to LGBTQ+ identities and coalitions of those "prohibited, stigmatized, and generally repressed" from heterosexuality, then a collective formed through a rejection of rigidity, austerity, and otherwise straightness, surely, does so through their proximities to being (and identifying) as queer themselves (Cohen 1997, 453). On the other hand, and more salient for our argument here, networked togetherness is queer insofar as it furthers José Esteban Muñoz's (2009) "understanding of queerness as collectivity" (11). We invoke queer as a relational process, a set of practices of kinship, intimacy, and care that refuses the strictures of wider relational norms that maintain the very dominant powers we contest in this chapter.

So let us be explicit: in the case of the university, networked togetherness can save lives. The university (whether online or not) is an institution "built upon the theft, the conquest, the negligence" (Harney and Moten 2013, 33) of rigidity and sameness (read: straightness), and therefore, it is indebted to concepts of property, professionalization, performance, and policing. Networked togetherness is a queer digital activism that, as Rianka Singh (2020) illuminates, is not one undertaken so that we can be heard but instead so we "can survive," seek and offer care, and succeed in systems designed to eradicate us.

Chat

Theresa to Alexis-Carlota (Direct Message)

Half of the time we don't even talk in these Zoom calls ⁻_(ツ)_/⁻ We just screech at each other😂😂😂
2:32 pm

Alexis-Carlota to Theresa (Direct Message)
looking forward to being the absolute worst with you this Friday on our residency call 🤪😹
2:33 pm

Many of us now know it well: click to join the meeting, await your microphone to synchronize, notice how many black boxes sit beside pixelated video streams of fellow attendees, turn on your own webcam, and watch as the red chat notifications appear along the taskbar. The Zoomscape has various digital affordances—technological conventions, buttons, profile settings, and icons— for interactivity and visuality. The computer screen, and more specifically the platform window's occupation of the screen, mediates an augmented virtual "face-to-face" interaction (at least through the webcam) with peers and instructors. This mediated communication, made possible through the visuality of the screen, provides a semblance of direct engagement, especially compared to mere recordings.

For the university, the assumed benefits of having synchronous online learning outweigh the ways in which the Zoomscape flattens visual cues, expects neurotypical learning, and produces its own rules of online etiquette, which are modeled after in-person university expectations. Platform functions like video sharing, Zoom breakout rooms, or instant polls encourage a level of engagement that rewards virtual participation and, in turn, demands participant performativity—a logic bound up in professionalism, productivity, attentiveness, and exchange. Put another way, these affordances do offer a means for interactivity but through methods that neglect the very real ways users might experience the Zoomscape differently, whether it be due to varying levels of surveillance or the need to wander (quite literally, away from their desks, or physically, to other applications on their screens). As easy as it is for the rigorous realities of the academy to infiltrate our digital and physical spaces in these ongoing pandemic times, Zoom as a platform provides affordances for alternative practices. If, as Lucas Graves (2007) defines, affordances are "features of a technology that make a certain action possible" (332), then we contend that Zoom's chat function is a feature of technology that affords multitasking, playing, wandering, and backchanneling in ways that permit precarious graduate student users (and others) the capacity to challenge normative academic structures and build community and kinship.

Zoom's chat function demonstrates how users can engage in a parallel chat (Sarkar et al 2021), wherein a text-based conversation flows concurrently with the audiovisual contributions, especially on the computer. Often positioned as two parallel window panels on the Zoom application interface, user screens become the site of *"multimodal multitasking* at its extreme—for example simultaneously engaging in a video chat while typing a text message (to the same or other individuals/groups) while checking and reposting a web link that is pertinent to the spoken conversation while reading and responding to an email, etc." (Graham 2019, 380–1). To some extent, this multitasking seems to align well with capitalist hyperproductivity as the user is participating across the platform's features. Yet, parallel chat has been described as distracting (Sarkar et al. 2021), and its encouraged multitasking has become a characteristic of Zoom fatigue (Riedl 2022), causing a decline in productivity overall.

As we see it, the Zoom chat's invitation into distracting multitasking is an affordance that lends itself to challenging what Elizabeth Freeman (2010) calls "chrononormativity, or the use of time to organize individual human bodies toward maximum productivity" (3). The chat function offers up opportunities for queer practices of relationality and connection in spite of the demands of professionalism and productivity in university online spaces. In similar observations of the classroom by Kristin Comeforo (2022), the Zoom chat was a "site where multiple and diverse voices would raise organic, and often parallel, discussions. Students used the Chat to share and excavate experiential knowledge, making it a powerful, community-centered knowledge (re)source" (1). These Zoom chat side conversations afford a mode of digital backchanneling, an intimate conversation that occurs simultaneously alongside a larger, more public conversation. Digital backchanneling can use a range of technologies to foster an informal discussion that takes place simultaneously alongside the live audiovisual discussion. The backchannel can be on- and off-the-focus platform. It can occur within the main chat or in direct messages. For us, this particularly took the form of precarious students having conversations with those deemed safe, such as other fellow students and trusted colleagues, in the Zoom chat through direct messages while being simultaneously present in a larger academic Zoom meeting hosted by the institution. The act of backchanneling, we suggest, supports the formation and sustenation of a collective culture that ultimately distracts and disorients expectations for the siloed productivity that defines the neoliberal university.

Chat

Theresa to <u>Alexis-Carlota</u> (Direct Message)
I love how this chat is: "I'm in crisis, here's a meme!" vibes
3:06 pm

                    Alexis-Carlota to <u>Theresa</u> (Direct Message)
                    I will burst in soon, unexpectedly, with a meme
                    3:07 pm

Theresa to <u>Alexis-Carlota</u> (Direct Message)
You're always radiating and beaming through the screens and stuff
3:07 pm

It was through the Zoom chat that we, the authors, first connected in 2020. Alongside our collaborators and friends, we cultivated a praxis of care and belonging. Our Zoom chats acknowledged our precarity as graduate students, queer scholars, and neurodivergent people expected to perform and produce while impending doom waited outside our doors in the form of an infectious virus. Zoom functioned alongside and aided feelings of isolation and loneliness. Through our relationships fostered by the Zoom chat, we cared for each other in various ways, sharing knowledge, supporting each other, and experiencing a collective frustration existing in a system not created for us. We engaged with and called for mentorship and learning beyond capitalist, colonial cultures of production enforced by the university and rejected the correlation of publications, research projects, grants, and CV points with self-worth. What first began as a platform thrust upon us by the neoliberal university to continue producing as "normal" was co-opted and queered as a space to challenge, reimagine, and organize against these expectations and the institutions that uphold them.

In contrast to the primarily academic conversations we were experiencing as graduate students within the Zoomscape (such as those relating to coursework, academic conference spaces, and other affiliations), the backchannel of the Zoom chat differed greatly in its affordances. These secondary, subcultural networks provided a space for subversion, offering what Angela McRobbie (1991) calls an "escape from the demands of [the] traditional" (26). It gave us the opportunity to speak back to the university's oppressive structures while conscious of our precarious statuses. As Singh (2020) argues, these environments of resistance place marginalized survival at the forefront. Our use of Zoom chat across these instances offers examples that challenge the neoliberal nature of academic cultures. Conscious of these colonial and capitalist realities, our question becomes: how do we subvert academic performance and put forth an alternative to the rigidity or sameness of the institution?

We turn briefly now to how humor, care, and fugitivity might enable and enact a networked togetherness, and in turn, we discuss how Zoom's chat function and its affordances like multitasking, backchanneling, and play offer precarious graduate students alternative avenues to do academia differently. Put another way: if, as we have proposed, networked togetherness is a set of digital, subversive, and collective practices, then we consider how the Zoom chat and its backchannels invite the precarious to develop a kind of togetherness through uses of humor and play that ultimately enact care and fugitivity. Such an approach conceives of humor as a critical tool that "emerge[s] in more immaterial, networked exchanges as social exchange, play, and critique" (Sundén and Paasonen 2020, 5). At the same time, turning to humor also recognizes the emergence of "pandemic humor," or the blending of "commonly understood childhood or popular cultural references with sarcastic humor," which "emphasizes references to COVID-19, including general issues, those who denied their vulnerability to COVID-19, and a coming apocalypse" (Aronson and Jaffal 2022, 523–6). Humor, shared in and between the precarious during precarious times, may seem quaint, but we argue that it can be an anti-colonial, feminist, queer tool that brings our focus back to practices of relation, care, and fugitivity that demand and dream for better landscapes both online and offline for learning in responsible and accountable ways. It is our contention that the Zoom chat offers a means to enact survival through community, facilitated by subcultural and subversive humor communicated through memes and emojis.

Internet memes often consist of jokes, satire, and sarcasm through intertextuality and "subcultural literacy: knowledge of the codes and norms developed in this meme-based subculture" (Shifman 2014, 115). In turn, memes create "collective identities through shared norms and values," which can support "collective coping in response to stressful situations where humour is used to mitigate such effects" (Ortiz et al. 2021, 169). Memes thus encourage a collective subculture, taking users into a level of communication that is necessarily relational. In the context of Covid-19, studies have explored how pandemic humor has become especially tied to memes and their circulation online, illustrating humorous, collective, and affective responses to rapid societal changes (see, e.g., Flecha Ortiz et al. 2021; Olah and Hempelmann 2021; and Aronson and Jaffal 2022).

In academic online spaces, the infamous "This is Fine" meme circulated through links, gifs, and references on social media and on Zoom. The "This is Fine" meme features two panels from K. C. Green's 2013 six panel webcomic "On Fire" from his series *Gunshow*. In this meme's two panels, a yellow anthropomorphic dog with black ears and a little yellow hat sits at a table with a cup of coffee. The room that the dog is in is visibly on fire, with flames and smoke rising to the top of the room. In the first comic panel, the dog's face lacks emotion despite the danger around them. In the second panel, they say "this is fine" with a smile when the situation is, presumably, not fine. "This is Fine" has been heavily circulated in various digital spaces beyond those with ties to the academy to depict the absurdity of perseverance amidst difficulty. The uptake of this meme (and its various edits) on the internet signals a shared experience of when "a situation becomes so terrible [that] our brains refuse to grapple with its severity" (Plante 2016), and therefore, it also tracks a shared affect: a mixture of dismay, discontentment, and disassociation.

Perhaps inevitably, the meme began to circulate across our Zoomscapes in the university. Links to the meme or "this is fine" comments were dropped in Zoom chat, privately or at large, affirming the exhaustion felt and the barriers experienced by professors, graduate students, and undergraduate students alike. However, our interests lie in how memes like "This is Fine" do more than engage with pandemic humor online due to the affordances facilitated by the Zoom chat. That is, in sharing memes or crafting jokes that require a particular subcultural context, the Zoom chat invites alternative modes of engagement that play with and puncture the formalities or rigidity of learning, meeting, and doing academia online. That is, the use of memes in Zoom chat backchannels interrupts the expectations of the propriety of the university online space. This humor distracts. It orients our focus elsewhere, drawing us to multitask and relate otherwise.

More saliently, memes might resonate with almost anyone, but the circulation of specific memes (like "This is Fine") in academic Zoomscapes prompts us to consider that there might be different levels of meme literacy at play, and therefore, subversive digital counterpublics, too. Much like how emojis can "function as a way to create and/or reinforce community" (Graham 2019, 393) in digital counterpublics, memes also register subculturally and counterpublicly for the already precarious in academia insofar as this humor points to not just the pandemic conditions of academia online but also to the ways in which some unequally feel the weight of what has historically been, and continues to be, a white, colonial, heteropatriarchal, ableist, and otherwise marginalizing institution. The circulation of this meme by queer, neurodivergent graduate students of color like us, for instance, speaks to a collective acknowledgment—being in the know—of the precarity we experience in the institution regardless of being online or offline. The Zoom chat thus affords a backchannel for such precarious students, operating as a collective digital counterpublic site to critique the institution and its structures, which requires a subcultural literacy to fully engage. This is an act of doing academia differently together—distracting, playing, laughing, and so on—through otherwise knowledges often discouraged in academic settings, pedagogies, and expectations and yet required to survive in such inhospitable spaces.

Such subcultural literacies, otherwise knowledges, collective work, and digital counterpractices exemplify and elicit a networked togetherness. In its relational provocations and practices, networked togetherness provokes us to question how "the present is so poisonous and insolvent" (Muñoz 2009, 51) particularly in the case of the university and the ways it sustains its white, colonial, capitalist, cisheteropatriarchal, able-bodied, and neurotypical logics both online and offline. That is, backchanneling in the Zoom chat can afford a networked "negligence of professionalization" (Harney and Moten 2013, 38) consisting of processes, practices, and pedagogies focused on survival in ways that sustain us by acknowledging the very crisis at hand and the care work necessary to survive it. Such fugitivity, like Harney and Moten's (2013) undercommons, that can "be understood as wary of critique, weary of it, and at the same time dedicated to the collectivity of its future, the collectivity that may come to be its future" (28). To recall Muñoz (2009) once more, this collectivity is insistent on where the relational can cut across overlapping capitalist and colonial expectations, moving us toward "the illumination of a horizon of existence" (47). And so, we observe and offer networked togetherness as a relational pedagogy and practice as a form of resistance, especially at times where all is not fine.

Chat

Alexis-Carlota to <u>Theresa</u> (Direct Message)
I'm very excited about meeting in person soon!!
4:59 pm

        THE HOST HAS ENDED THE ZOOM MEETING

Acknowledgments

The authors would like to thank Andrea Zeffiro for her support of this research project and our pedagogical experiments. We are grateful, too, to Linzey Corridon and Maddie Brockbank for their involvement in practicing a networked togetherness on Zoom (and inspiring our chapter epigraphs).

References

Antwi, Phanuel. 2018. "On Labour, Embodiment, and Debt in the Academy." *A/b: Auto/ Biography Studies* 33, no. 2: 301–26. https://doi.org/doi:10.1080/08989575.2018 .1445577.

Aratani, Lauren. 2020. "'Zoom University': Is College Worth the Cost Without the In-Person Experience?" *The Guardian*, October 6, 2020. https://www.theguardian .com/world/2020/oct/06/zoom-university-college-cost-students-in-person -experience.

Aronson, Pamela and Islam Jaffal. 2022. "Zoom Memes for Self-Quaranteens: Generational Humor, Identity, and Conflict During the Pandemic." *Emerging Adulthood* 10, no. 2: 519–33. https://doi.org/10.1177/21676968211058513.

Batac, Monica Anne. 2022. "'Failing' and Finding a Filipina Diasporic Scholarly 'Home': A De/Colonizing Autoethnography." *Qualitative Inquiry* 28, no. 1: 62–9. https://doi .org/10.1177/10778004211006705.

boyd, danah. 2011. "Social Network Sites as Networked Publics: Affordances, Dynamics, and Implications." In *A Networked Self: Identity, Community and Culture on Social Network Sites*, edited by Zizi Papacharissi, 39–58. New York: Routledge.

Breen, Marcus. 2021. "Coronavirus Pedagogy in the Zoomscape: Pinhole Intimacy Culture Meets Conscientization." *Lateral* 10, no. 2. https://doi.org/10.25158/L10 .2.16.

Chen, Mel Y. 2012. *Animacies: Biopolitics, Racial Mattering, and Queer Affect*. Durham: Duke University Press.

Chun, Wendy Hui Kyong. 2017. *Updating to Remain the Same: Habitual New Media*. Cambridge: The MIT Press.

Comeforo, Kristin. 2022. "Witnessing with Cameras off: Feminist Pedagogy and the Zoom Classroom." *Feminist Pedagogy* 2, no. 3: 1–4. https://digitalcommons.calpoly .edu/feministpedagogy/vol2/iss3/7.

DiNucci, Darcy. 1999. "Design & new Media: Fragmented Future-Web Development Faces a Process of Mitosis, Mutation, and Natural Selection." *Print Magazine* 53: 32–5.

Duguay, Stefanie, Anne-Marie Trépanier, and Alex Chartrand. 2022. "The Hottest New Queer Club: Investigating Club Quarantine's off-Label Queer Use of Zoom during the COVID-19 Pandemic." *Information, Communication & Society*, 1–17. https://doi .org/10.1080/1369118X.2022.2077655.

Eglash, Ron. 2004. "Appropriating Technology: An Introduction." In *Appropriating Technology: Vernacular Science and Social Power*, edited by Ron Eglash, Jennifer L. Croissant, Giovanna Di Chiro, and Rayvon Fouché, vii–xxi. Minneapolis: University of Minnesota Press.

Finley, Chris. 2020. "Building Maroon Intellectual Communities." In *Otherwise Worlds: Against Settler Colonialism and Anti-Blackness*, edited by Tiffany Lethabo King, Jenell Navarro, and Andrea Smith, 362–70. Durham: Duke University Press.

Flecha Ortiz, José A., Maria A Santos Corrada, Evelyn Lopez, and Virgin Dones. 2021. "Analysis of the Use of Memes as an Exponent of Collective Coping during COVID-19 in Puerto Rico." *Media International Australia* 178, no. 1: 168–81. https://doi.org /10.1177/1329878X20966379.

Freeman, Elizabeth. 2010. *Time Binds: Queer Temporalities, Queer Histories*. Durham: Duke University Press.

Gilmore, Ruth Wilson. 2007. *Golden Gulag: Prisons, Surplus, Crisis, and Opposition in Globalizing California*. Berkeley: University of California Press.

Graham, Sage L. 2019. "A Wink and a Nod: The Role of Emojis in Forming Digital Communities." *Multilingua* 38, no. 4: 377–400. https://doi.org/10.1515/multi-2018 -0037.

Grandinetti, Justin. 2022. "'From the Classroom to the Cloud': Zoom and the Platformization of Higher Education." *First Monday* 27, no. 2. https://doi.org/10 .5210/fm.v27i2.11655.

Graves, Lucas. 2007. "The Affordances of Blogging: A Case Study in Culture and Technological Effects." *Journal of Communication Inquiry* 31, no. 4: 331–46. https:// doi.org/10.1177/0196859907305446.

Harney, Stefano and Fred Moten. 2013. *The Undercommons: Fugitive Planning & Black Study*. New York: Minor Compositions.

Hartman, Saidiya. 2020. "The Death Toll." In *The Quarantine Files: Thinkers in Self-Isolation. Los Angeles Review of Books*, April 14, 2020. https://lareviewofbooks.org/ article/quarantine-files-thinkers-self-isolation/#_ftn15.

Hill, Marc Lamont. 2018 "'Thank You, Black Twitter': State Violence, Digital Counterpublics, and Pedagogies of Resistance." *Urban Education* 53, no. 2: 286–302. https://doi.org/10.1177/0042085917747124.

Hobart, Hiʻilei Julia Kawehipuaakahaopulani and Tamara Kneese. 2020. "Radical Care." *Social Text* 38, no. 1: 1–16. https://doi.org/10.1215/01642472-7971067.

Hunt, Sarah and Cindy Holmes. 2015. "Everyday Decolonization: Living a Decolonizing Queer Politics." *Journal of Lesbian Studies* 19, no. 2: 154–72. https://doi.org/10.1080 /10894160.2015.970975.

Jenkins, Henry. 2016. "Youth Voice, Media, and Political Engagement: Introducing the Core Concepts." In *By Any Media Necessary: The New Youth Activism*, edited by Henry Jenkins, Sangita Shresthova, Liana Gamber-Thompson, Neta Kligler-Vilenchik, and Arely M Zimmerman. New York: New York University Press.

Jiwani, Yasmin. 2020. "The Intersecting Violence(s) of a Pandemic." *TOPIA: Canadian Journal of Cultural Studies* 41: 115–22. https://doi.org/10.3138/topia-014.

Kenney, Theresa N., Emily Goodwin, Alexis-Carlota Cochrane, Linzey Corridon, Maddie Brockbank, and Sarah Paust. 2023. "Constellations of Community, Care, and

Knowledge: A Collection of Vignettes from Pandemic Times." *IDEAH*, June 6, 2023. https://doi.org/10.21428/f1f23564.760119d3.

Luka, Mary Elizabeth, Alison Harvey, Mél Hogan, Tamara Shepherd, and Andrea Zeffiro. 2015. "Scholarship as Cultural Production in the Neoliberal University: Working Within and Against 'Deliverables.'" *Studies in Social Justice* 9, no. 2: 176–96. https://doi.org/10.26522/ssj.v9i2.1138.

McRobbie, Angela. 1991. *Feminism and Youth Culture: From "Jackie" to "Just Seventeen"*. Hampshire: Macmillan.

Muñoz, José Esteban. 2009. *Cruising Utopia: The Then and There of Queer Futurity*. New York: New York University Press.

Nakamura, Lisa. 2002. *Cybertypes: Race, Ethnicity, and Identity on the Internet*. New York: Routledge.

Olah, Andrew R. and Christian F. Hempelmann. 2021. "Humor in the Age of Coronavirus: A Recapitulation and a Call to Action." *HUMOR* 34, no. 2: 329–38. https://doi.org/10.1515/humor-2021-0032.

Pathak, Dev Nath. 2022. "Idea of Pandemic-Pedagogy: Reflexive Rumination on Teaching and Learning Practices." *Higher Education for the Future* 9, no. 1: 62–74. https://doi.org/10.1177/23476311211046184.

Plante, Chris. 2016. "This Is Fine Creator Explains the Timelessness of His Meme." *The Verge*. https://www.theverge.com/2016/5/5/11592622/this-is-fine-meme-comic.

Renninger, Bryce J. 2015. "'Where I Can Be Myself … Where I Can Speak My Mind' : Networked Counterpublics in a Polymedia Environment." *New Media & Society* 17, no. 9: 1513–29. https://doi.org/10.1177/1461444814530095.

Riedl, René. 2022. "On the Stress Potential of Videoconferencing: Definition and Root Causes of Zoom Fatigue." *Electronic Markets* 32, no. 1: 153–77. https://doi.org/10.1007/s12525-021-00501-3.

Risam, Roopika. 2018. "What Passes for Human? Undermining the Universal Subject in Digital Humanities Praxis" In *Bodies of Information: Intersectional Feminism and Digital Humanities*, edited by Elizabeth Losh and Jacqueline Wernimont, 72–92. Minneapolis: University of Minnesota Press.

Sarkar, Advait, Sean Rintel, Damian Borowiec, Rachel Bergmann, Sharon Gillett, Danielle Bragg, Nancy Baym, and Abigail Sellen. 2021. "The Promise and Peril of Parallel Chat in Video Meetings for Work." In *Extended Abstracts of the 2021 CHI Conference on Human Factors in Computing Systems*, 1–8. Yokohama: ACM. https://doi.org/10.1145/3411763.3451793.

Shaw, Adrienne and Katherine Sender. 2016. "Queer Technologies: Affordances, Affect, Ambivalence." *Critical Studies in Media Communication* 33, no. 1: 1–5. https://doi.org/10.1080/15295036.2015.1129429.

Shew, Ashley. 2020. "Ableism, Technoableism, and Future AI." *IEEE Technology and Society Magazine* 39, no. 1: 40–85. https://doi.org/10.1109/MTS.2020.2967492.

Shifman, Limor. 2014. *Memes in Digital Culture*. Cambridge: The MIT Press.

Singh, Rianka. 2020. "Resistance in a Minor Key." *First Monday 25*, no. 5. https://doi.org/10.5210/fm.v25i5.10631.

Suarez-Villa, Luis. 2009. *Technocapitalism: A Critical Perspective on Technological Innovation and Corporatism.* Philadelphia: Temple University Press.

Sundén, Jenny and Susanna Paasonen. 2020. *Who's Laughing Now? Feminist Tactics in Social Media.* Cambridge: The MIT Press.

Warner, Michael. 2002. "Publics and Counterpublics." *Public Culture* 14, no. 1: 49–90. https://doi.org/10.1215/08992363-14-1-49.

Yusoff, Kathryn. 2018. *A Billion Black Anthropocenes or None.* Minneapolis: University of Minnesota Press.

Zeffiro, Andrea. 2021. "From Data Ethics to Data Justice in/as Pedagogy (Dispatch)." *Studies in Social Justice* 15, no. 3: 450–57. https://doi.org/10.26522/ssj.v15i3.2546.

Zuroski, Eugenia. 2020. "'Where Do You Know From?': An Exercise in Placing Ourselves Together in the Classroom." *MAI Feminism* 5. https://maifeminism.com/where-do-you-know-from-an-exercise-in-placing-ourselves-together-in-the-classroom/.

9

Stage Directions and Snarky Comments

Shadow Networks in Zoom Meetings Throughout the Covid-19 Pandemic

Heather J. Carmack, Mayo Clinic, Heather M. Stassen,
Daemen University, Tennley A. Vik, Independent Researcher, and
Jocelyn M. DeGroot, Southern Illinois University Edwardsville

Within an organization (a *legitimate system*), workers sometimes spontaneously and informally establish their own communication network: a shadow network (Stacey 1996). Shaw (1997) described a shadow system as a "complex web of interactions in which social, covert political and psycho-dynamic systems coexist in tension with the legitimate system" (235). Shadow networks have long been studied for their role in creativity and their ability to enact change (e.g., Shaw 1997). Transformations of and increases in the use of Zoom and other videoconferencing tools (e.g., Microsoft Teams, Google Meet) because of the COVID-19 pandemic highlight the need to examine shadow networks and the unique and covert ways employees can communicate within otherwise legitimate platform systems. Employees can now text, email, or private message people who are also on Zoom during a call, essentially participating in technology-based note-passing. Additionally, the unique nature of shadow network communication during Zoom calls means that workers can engage in multiple shadow interactions simultaneously, communicating with multiple other employees and niche groups about the main Zoom meeting.

In this project, we examine rhetorically constituted narratives of four authors from different institutions about their experiences engaging in backstage communication via their shadow networks during the first years of the Covid-19 pandemic. Organizational leadership typically communicates with workers through front-facing structures and channels in legitimate organizational networks (Ellingson 2004, building on the work of Goffman 1959), whereas shadow networks represent a form of backstage communication that happens in workplaces. During the first years of the Covid-19 pandemic, faculty, as organizational members, may have used Zoom as both a front stage (Zoom faculty meetings) and backstage (Zoom's private message function) to simultaneously communicate in legitimate and shadow networks. Shadow networks as communicative systems can provide opportunities for coalition formation and organizing outside the formal network of the Zoom meeting; however, they are frequently perceived by administrators and supervisors as challenging or disruptive to the legitimate system. Often, both in the digital realm and in face-to-face interactions, management and administrators attempt to silence the shadow network(s). We argue that these forms of backstage communication (e.g., texting others on the meeting or private Zoom chatting) are taking place within and simultaneously creating shadow networks. Weaving together literature about shadow networks and our experiences with shadow systems in higher education (e.g., virtual faculty meetings, university webinars, Faculty Senate meetings, and meetings with administrators at all levels of our universities) and the innumerable Zoom calls during the Covid-19 pandemic, we interrogate the ways we used Zoom and other mediated channels to communicate in the backstage with our shadow networks.

Entering the Zoom Front Stage during Covid-19

When the Covid-19 pandemic hit the United States in early 2020, many colleges and universities were preparing for spring break. For some faculty, that would be the last they were in their offices, classrooms, or campuses for months or years. Meetings transitioned completely online, with faculty learning how to navigate Zoom or Microsoft Teams. Like many faculty across the United States and the world, we found ourselves flung into the world of Zoom meetings, forced to learn the new language of Zoom, Teams, Slack, and other work platforms. Our experiences differed wildly, in part because of how our universities handled the pivot to online and in part because of the campus roles we inhabited during the beginning of the pandemic. Table 9.1 explains our academic roles as well as the Covid-19 activities and task forces that contributed to our involvement with shadow networks during work from home (WFH) mandates. Two important points of note: First, the authors have decided to use their real names next to their narratives. What is presented is our narrative understanding of our experiences. This was a challenging decision because we opted to share criticism and negative

Table 9.1 Author Roles and Activities during WFH

Author	Faculty Role	Administrative Role(s)	Covid-19 Activities
Heather Carmack	Associate Professor	MA Graduate Coordinator	Taught mostly in-person classes; online classes only for online MA program (*Note*: This is not the author's current role or employer)
Heather Stassen	Professor	Faculty Chair; Program Director; General Education Coordinator	Coordinated all-college faculty meetings and information sharing; member of the Campus Reopening Committee; daily meetings to prepare for fall reopening; hosted monthly all-faculty meetings to share administrative reports; facilitated faculty discussion about safety protocols and concerns (*Note*: This is not the author's current role or employer)
Tennley Vik	Assistant Professor		Taught online exclusively, was largely sheltered from on-campus activities and service. All meetings and classes took place via Zoom for 2021. Qualified for online only due to preexisting condition
Jocelyn "Josie" DeGroot	Professor	Assistant Department Head; Faculty Senate President	Co-chair of campus-wide Covid-19 pandemic planning task force; met with faculty, staff, and students on Zoom; hosted multiple webinars to gather information and disseminate decisions

experiences. Second, both Heather C. and Heather S. changed institutions after the start of the pandemic; their stories do not report their experiences at their current institutions.

Shadow Networks

Organizations function through the creation and maintenance of communication practices, rules, and hierarchies (Monge and Contractor 2001). However, the complexity of modern organizations often leads to the emergence of shadow networks or shadow structures (Shaw 1997; Stacey 1996). Shadow network communication can utilize multiple communication formats and technologies, ranging from informal hallway meetings to emails. The use of new technologies, especially during Covid-19, allows shadow network communication to occur on multiple devices simultaneously or to happen during formal communication interactions. Pre-Covid-19, it was common to text during in-person meetings. During Covid-19, the introduction of Zoom as a replacement for in-person meetings meant organizational members now had multiple media options to communicate privately during meetings; members could send a text message, but they could also direct message on a social media platform or send private Zoom messages while in a public Zoom meeting. These dyadic side conversations—private conversations using technology to communicate during public meetings—help shadow networks to flourish when in-person meetings were not possible.

Shadow networks have several key characteristics. First, shadow networks are emergent, formed through impromptu and improvisational communication and practices (Salem 2021). Second, they do not have a centralized or formal structure; however, they do have governable patterns of behaviors and communication expectations (Shaw 1997). Third, because shadow networks are self-organized, they are very fluid and can form and disband quickly (Kadushin and Brimm 1990). This fluidity and lack of strict communicative boundaries also allow organizational members to participate in multiple shadow networks. Fourth, and as our narratives later in the chapter showcase, shadow networks serve a purpose (Shaw 1997); they are strategic communication networks that provide "ways of by-passing or ignoring the formal structure so organization members can get things done" (Kadushin and Brimm 1990, 5).

There are drawbacks and benefits to the use of shadow networks. Shadow networks are, at their core, a challenge to the legitimate organization (Behrens 2009; Shaw 1997; Stacey 1996). Shadow networks are almost always framed by management as *bad* communication structures because they are perceived as undermining official communication structures, corrupting formal processes, and taking away from organizational members' work focus (Behrens 2009). Instead of working through official communication channels with the approval of organizational leadership, shadow network members use alternative structures. Kadushin and Brimm (1990) argued that this creates a double bind because leadership wants organizational members to communicate with each other but to do so without excluding or bypassing their authority.

Perhaps more problematic, the communication within shadow systems may be viewed as destructive or constructive, depending on one's perception of the in-group/out-group dynamic created by these informal communication networks (Kadushin and Brimm 1990). A common by-product of shadow networks is the creation of factions (Behrens 2009). Not every organizational member can be part of a shadow network; otherwise, it would be part of the formal communication structure. That means some organizational members will be excluded. These feelings of exclusion can lead organizational members to feel they are not part of the organization. Additionally, because shadow networks often have an agenda or are the result of the politics of the organization, this can lead to "factional warfare" (Behrens 2009, 128) and create hostility between members of the shadow network and those outside the shadow network, especially leadership. Supervisors typically feel challenged by these groups, seeing shadow communication as a direct threat to their leadership. When leadership attempts to discipline shadow network members, unintended consequences can occur (Stacey 1996), such as continued use of shadow networks or counterstriking. For example, faculty operating within a shadow network may engage in malicious compliance or reduce meeting availability when the network is disciplined or infiltrated by an administrator.

Shadow networks can, however, create space for positive communication and change in an organization. Shadow networks can be the result of organizational members' frustrations with the lack of ingenuity and freedom in their organization. The emergent and impromptu genesis of shadow networks means that the organizational members within the networks are capable of spontaneity, novelty, and thought agility (Shaw 1997). This creativity is especially important when the shadow network is formed to help organizational members navigate organizational dysfunction or leadership deficiencies (Behrens 2009). Additionally, shadow networks can create space for emergent and participatory organizing practices that can facilitate real organizational change (Shaw 1997; Stacey 1996). The analysis of our experiences with shadow networks highlights how shadow networks can be used to challenge problematic organizational practices and create space for consensus building and room for marginalized voices.

Shadow Network Capable

Heather S.: The messages came in from all directions—text message, Facebook messenger, Zoom's direct chat function, Teams chat, and email. Someone even called me via Microsoft Teams during a Zoom meeting I was facilitating, and they were attending! I half expected owls to start sliding notes into the crevices of my home, Harry Potter style.

The Covid-19 pandemic created an exigent situation, where technology served not merely as a tool but also as a way to communicate about the experience of using technology. Societies place meaning onto the use of technologies, and because humans use them to connect with each other, affordances are often considered social (Fox and McEwan 2017). Channels used to communicate are socially embedded and constructed; how we use and understand technology changes and differs depending on context, understanding of technology, past experiences, current societal appreciation for or hatred of technology, and the standards and expectations applied to its use at any given time (Fox and McEwan 2017; Mascheroni and Vincent 2016; Schouten et al. 2007). As Heather S.'s narrative illustrated, she fielded messages from multiple channels, including people using multiple channels simultaneously to reach out for information—including colleagues asking about how to use the technology. During meetings discussing reopening plans, faculty frequently utilized channels outside of the legitimate Zoom meeting to pose questions to senior leadership, mediated through senior faculty and faculty leadership, with the expectation that they would be shielded. Faculty who may have otherwise felt silenced in a face-to-face meeting (e.g., untenured faculty, contingent faculty) utilized accessible channels to reach out to Heather S. in her capacity as a tenured professor and faculty chair to question leadership's decision-making, particularly in reference to safety protocols. Heather S. was in a perceived position of representation as the elected chair of the faculty, but she was still able to engage in shadow communication because decisions were being made by larger organizational forces. However, not all faculty had this level of access—those with network associations within the shadow networks Heather S. haunted were able to communicate with her in these media (e.g., group chats) while others in out-groups were excluded. Ultimately, this led to certain shadow networks controlling the faculty narrative posited to senior leadership.

The inclusion of multiple channels to facilitate communication during the first years of the pandemic set the stage for shadow networks to emerge in force, as organizational members turned to different modalities to communicate. The introduction of WFH created a workspace where organizational members were able to, or were required to, use multiple platforms to perform work. As

Josie explained, faculty would turn to multiple platforms simultaneously to communicate:

> Josie: We talked via text, private message, and Facebook messenger. This occurred during in-person faculty meetings, but it was especially evident during Zoom meetings. Pleasantries are exchanged privately because we didn't run into each other in the hall to have informal conversations and get caught up, and we didn't want to interrupt the flow of the Zoom meeting. In a face-to-face meeting, more than one conversation can happen at a time. The agenda topic might be in formal discussion, and to those it didn't apply to (or interest), those people might have a quiet side conversation. Because we were texting or private messaging, we were able to vent about a certain person or action and just generally be snarkier than we would be in an in-person meeting.

Communicating in private channels or venting does not inherently make the communication part of a shadow network or shadow communication. However, the affordances provided by using multiple channels created space for shadow networks to emerge. Heather C.'s story below illustrates how frustrations coupled with multiple technologies allow for shadow networks to emerge.

> Heather C.: I remember one very dysfunctional faculty meeting. We hadn't received the agenda or the Zoom invite from the department head—one of the faculty emailed about 45 minutes before the meeting to see if we were even having it. At the meeting, people arrived late (triggering the doorbell sound on Zoom), the department head was directing people in the meeting to email documents to the faculty in the meeting (so email chimes—mine because my email was open and others' email notifications I could hear on the Zoom call), survey chimes were sounding because the department head was launching surveys to collect votes on the documents we just received in the email (and hadn't reviewed), faculty were sending messages in the Zoom chat (both public and private), and faculty were side texting in a group text chat about the dysfunction of the meeting. Not surprisingly, there was a lot of confusion about what was happening, and what faculty were supposed to talk about and vote on. I eventually just turned off my camera, closed my email, and voted "abstain" when a poll would pop up. Most of the meeting was spent messaging in the group chat and private Zoom chat about how bad the meeting was and how we could fix it.

As Heather C.'s story highlights, an already existing shadow network relied on their previous communication channels (group text) but also commandeered the legitimate organization's communication channel (Zoom) to engage in shadow communication. In Heather C.'s case, the shadow communication focused not only on the dysfunction but also how members of the shadow network, if given the opportunity, would improve the meeting. In this case, however, the power imbalance between the department head and members of the shadow network meant that recommendations for improvement were not solicited or given. Instead, the shadow network used the digital space to voice concerns about the department head's turbulent and chaotic organizational style in a way that did not challenge leadership.

Stage Directions, Backstage Communication, and Grapevine Communication

Josie: I was in my second semester as Faculty Senate President when the pandemic led to Illinois's stay-at-home order. We held meetings online. I would send a Zoom PM to council chairs to double-check on their report or other meeting-based directions or nudge someone to call for a vote. Later we were told we could not use the private messaging because of the Illinois Open Meetings Act that requires all communication and deliberation during a qualifying meeting to be public. During the COVID-planning webinars, I would also text my committee co-chair if he was talking too long or to let him know the next topic we would discuss.

Shadow networks thrive on nondirectional and unbounded communication practices (Shaw 1997). Informal communication structures, often referred to as grapevine communication, relay information to individuals or groups outside the traditional channels and workflows of the legitimate organization (Crampton et al. 1998). Grapevine communication is often associated with gossip or rumor (Burke and Wise 2003; Kast 2019); however, that does not mean the information provided in this structure is false or malicious. Rather, informal communication structures can assist in disseminating information to organizational members, present information in a clearer format, promote sensemaking, and help organizational members develop connection (Robinson and Thelen 2018). This was especially true during the early stages of the pandemic, when individuals relied on different technologies and media to communicate (Sarai and Gotora 2021). Josie used informal communication as a way to provide *stage directions*, to help focus meetings, and to assist others in their new pandemic-related work.

While backstage is traditionally understood as "a place, relative to a given performance, where the impression fostered by the performance is knowingly contradicted" (Goffman 1959, 112), this only holds partially true in shadow networks and organizational settings. Backstage communication in organizations is still professional communication; it just takes place in back-facing spaces (such as hallways, breakrooms, or behind office doors). As Ellingson (2003, 2004) discovered, backstage workplace communication is often spontaneous and impromptu, as organizational members exit the front stage to accomplish work in the backstage. This is the space where organizational members can share important information, build relationships, and identify plans of action (Wittenberg-Lyles et al. 2009). Scholars examining backstage communication in organizations have been clear that backstage communication is not intended to be secretive or *bad* communication; rather it is meant to be private communication that may not involve all the frontstage actors (e.g., Ellingson 2003, 2004; Tracy 2000). Although we did notice some more casual tones in communication (including *snark*, as Josie notes), shadow networks still often maintained a professional tone throughout the pandemic but used backstage channels (e.g., Zoom private chat, text message) to communicate. Undoubtedly, tighter shadow networks both in terms of size and perceived level of trust allowed for less formal presentations of self to take place. The ability to shift a camera angle, turn off a camera momentarily, or easily send a direct message made for greater ease of shadow communication than sneaking in a quick discussion during a face-to-face meeting.

Importantly, there are still expectations and rules for shadow communication, which may or may not be formally communicated to organizational members. For organizational members working in shadow networks, these expectations may be assumed, creating room for disagreement.

> *Tennley: Before Zoom meetings often I would talk to other friends and coworkers about what was expected in the meetings, and also debrief after a meeting. Of course, what occurs in the meeting is the front stage (legitimate) communication that occurs in front of everyone in the meeting, but the pre-meeting phone calls as well as a debrief call after a Zoom meeting occur in the backstage (shadow communication). The content of these conversations varied dramatically based on who is communicating (the actors) and what the context of the interaction is (the script).*

When explaining her shadow communication, Tennley highlighted the ways that backstage expectations still dictated the form and content of communication. In a shadow network, the challenge may be to determine what to communicate and to whom to communicate it.

Heather S.: The pre-pandemic hallway conversations often heard in hushed tones and murmurs, the ones that can create a consistent hum through an academic corridor before and after a meeting, pivoted to the remote environment. The notes slipped from untenured faculty to mentors on the other side of the academic gauntlet went from Post-its and scraps off the bottom of a meeting agenda to individual chats and texts through myriad technological tools. The difference, from my perspective, was the synchronicity. The ability for folks to communicate, in real-time and without notice, to and with those who were safer (e.g., the tenured, the Fulls) allowed questions to emerge and challenge administrators.

Heather S.'s narrative also highlights an important and often ignored element of shadow networks: they are meant to be safe spaces for organizational members to communicate with each other, without fear of reprisal. Shadow networks emerge as the result of frustration with the current organizational system *and* frustration with the inability to speak up or challenge those systems. Shadow networks' move to technologically mediated channels meant individuals could not only dialogue in real time about issues but also feel safe in that communication space.

Leveraging Shadow Networks for Coalition Building and Faction Warfare

Tennley: I reached out to tenured faculty members for support and coalition building. Specifically, I did not feel comfortable expressing a concern in a large college-wide faculty meeting. I chose to reach out to a tenured faculty in another department so he could be my voice to share my concerns in the meeting. This person was able to keep my identity private. He also made sure to say that he was voicing the concerns of faculty outside of his own home department. I appreciated the anonymity that this other faculty member was able to provide for me, as I was afraid of retaliation from inside and outside my home department.

The primary goal of shadow networks is to find alternative means of accomplishing goals. Although shadow networks can create problems for organizational members, they are meant to be a communicative space for members to come together outside the confines of the legitimate organizational structure (Salem 2021; Shaw 1997). As such, one of the major benefits of shadow networks is the ability for members to engage in coalition building. Heather C. explained how the shadow network was used in her former department to whip votes for departmental change:

Heather C.: Definitely coalition building among some of the junior and senior faculty to get them to vote in block on certain agenda items. There were several items we voted on that junior faculty cared about but were worried about voicing concerns or support for. We texted before and during meetings so that we could present a united front and vote together to support.

Tennley's and Heather C.'s experiences represent the way shadow networks can be used to virtually "pass notes" to accomplish goals. For Tennley, shadow communication helped her find a messenger when she could not voice her concerns that protected her from retaliation. In Heather C.'s case, the shadow network was used for its truest purpose: to accomplish change. Deploying the shadow network showed how the network was able to use shadow communication to champion their cause.

However, using shadow communication to coalition build in meetings does have an unintended consequence in that it creates in-group/out-group dynamics. The point of coalition building is to bring together a group to counter another group, which means, ultimately, people are left out of the shadow network. It is not necessarily a problem to do this when individuals in the shadow network do so to create safe spaces or use it to build a group to accomplish their goals. However, Behrens (2009) argues that shadow networks can give rise to factions in an organization and lead to "factional warfare" (128). Heather C. and Tennley experienced this firsthand. In these cases, grapevine communication became a primary communication channel, either as the result of warfare (Tennley) or to further entrench warfare (Heather C.).

Tennley: Several of my colleagues stopped attending faculty meetings as a form of protest. Because of this choice, the information filtered to these colleagues about faculty meetings was primarily through grapevine communication. These colleagues were open to receiving this information, but it was filtered to them largely through two other faculty members (not me), and that information provided the two faculty not attending meetings with a snapshot of what happened in the meeting, but not the meeting in its entirety. This form of grapevine communication is especially important when faculty members miss a Zoom meeting for any reason, but especially because these two faculty members were expressing dissent through not attending meetings.

Heather C.: Pre-pandemic, informal communication was how information was filtered through the department, which was already heavily criticized by faculty who felt excluded and "out of the loop." This only got worse during the pandemic.

The new department head identified an "inner circle" (that was what the department head actually called them) and those four people filtered information down to whomever they chose. Instead of sending out department-wide emails about important department business, faculty would get information second or third-hand through text messages from other faculty. Whether you heard directly from someone in the inner circle depended on whether you were considered an insider or outsider in the department. I was firmly in the outsider group, which meant that I often received information thirdhand or not at all.

Tennley's and Heather C.'s competing narratives, although both related to faction warfare, underscore issues of power within organizational groups. Both warfare engagements hinge on insider/outsider dynamics, either self-imposed or imposed by others. Inclusion and exclusion, then, can be a driver of shadow networks or the result of shadow networks. Shadow networks differ from workplace cliques in that they are engaging in organizational business, not just gossip or socialization. The use of digital tools with ease during the pandemic both reified existing networks and created new ones and, because of their occurrence in a digital space, were less obvious than colleagues frequenting certain offices with hushed tones. One of the disadvantages of shadow networks is division, and in both stories, division was the centerpiece that led to or reinforced shadow networks. However, it is important to note the (dis)empowerment of organizational members in these stories. In Tennley's case, the colleagues who relied on grapevine communication did so as an empowerment mechanism— the shadow network allowed them to express dissent and facilitate their protest. Heather C.'s case is very different. The faction warfare created by the "inner circle" further disenfranchised her. Although the outsider group could engage in shadow communication, it was still incomplete because the insider group controlled the communication.

Disciplining, Surveilling, and Counterstriking Shadow Networks

Heather C.: During Zoom meetings, I would Zoom private message another faculty member—it was a running gag related to the problems of my WFH setup. It was usually just little jokes. When the department head found out we were using the private message function (and that other faculty were engaging in similar practices), they turned off the private message function in Zoom. That faculty member actually attempted to continue the jokes, only in the frontstage during a meeting. When that happened, the department chair immediately reprimanded him about how inappropriate it was and suggested it should be noted in his personnel file. This forced more shadow communication to take place via text message and debrief phone calls.

Administrators and leadership typically distrust shadow networks and frame this communication as destructive to organizational morale and usurping their authority. Framing shadow networks as disruptive seeks to justify surveillance and discipline. Foucault (1995) argued that power is exercised to discipline individuals when they engage in behaviors those in power deem inappropriate or unallowable. The legitimate organization would argue that this is in the best interest of all organizational members—that the formal structures of the organization bestow legitimate authority on managers and leadership (Sewell and Barker 2006). Shadow networks operate outside the surveillance space, making it difficult to control. To the legitimate organization, this means that workplace activities cannot be scrutinized.

Platforms like Zoom afford organizational members space to engage in shadow communication; however, because these platforms are often controlled by leadership, leadership has the ability to enact controls and tools that are baked into platforms such as Zoom under the auspices of efficiency (e.g., controlled sharing, the host's ability to mute others). Heather C.'s story identifies the easiest way for leadership to attempt to discipline and control shadow network communication—turning off access to the channels they use during formal interactions (a much easier task than hushing hallway conversations pre-pandemic). Not surprisingly, however, this drives shadow network members to use other channels to communicate, moving deeper into the shadow network and away from the legitimate network. This comes at a cost, however, because these are channels leadership cannot control, which creates additional frustration, surveillance, and discipline. Josie's experience with leadership control highlights important elements of shadow networks (e.g., grapevine and backstage communication; information sharing) as well as shows how shadow networks organically grow and adapt as a result of authority surveillance and discipline:

Josie: When watching webinars or Zoom sessions conducted by university administrators, I would use Zoom private messaging to talk with fellow faculty and staff members to critique or clarify what was being communicated. In these private conversations, I began learning more non-publicly available (yet, not confidential) information about the University as a whole. After a few meetings, the administration turned off private messaging in the webinars. This essentially blocked [faculty from] being able to privately discuss the decisions they were making. As a result, I found myself texting the people whose phone numbers I did know. They then looped me into group texting conversations with people whose phone numbers I did not originally have. Further, this blocked means of communicating led to feelings of distrust. What was the admin hiding? Why don't they want us talking to each other? Zoom webinar settings meant you could only see and hear the presenters. Participants would watch and listen to the presentation. A question-and-answer box was available for the presenters; however, there is a setting to make those questions private to the presenters only until the question was asked. This means that some questions were never answered, and no one knew any better. Because the admin/CORE team used the option to hide the questions, they further created distrust. This was especially obvious (and negatively viewed), as the university community was used to seeing the Academic Continuity Task Force (ACTF) webinars that were fully transparent, allowing chatting and posting the questions for all to see.

Josie's story showcases a common discussion point in shadow networks, especially when they discipline and control communication: What are they hiding? Shadow networks are frequently born out of frustration with the legitimate organization, and when the legitimate organization attempts to control communication channels, it reinforces the shadow network members' beliefs that the organization is problematic. Failure by management and administrators to allow for shadow networks can, and often does, further exacerbate skepticism and distrust. In Josie's case, it was a real concern that leadership was hiding information related to the Covid-19 pandemic, which further fostered feelings of frustration and distrust. Additionally, it is important to note that the leadership not only blocked one shadow network channel (Zoom private messenger) but also attempted to control communication in the legitimate network by using a question-and-answer system that only they could see, allowing them to determine what information was disseminated, what questions were answered, and what knowledge was deemed relevant for the groups.

Shadow networks can identify workarounds or alternative means to counter discipline and surveillance—in essence, they can counterstrike anticipated discipline. Although leadership's approach to surveillance and discipline of emerging technologies may be to simply shut down channels, organizational members often identify workarounds to using those technologies (Manley and Williams 2022); however, that is not without pushback and consequence.

Heather C.'s narrative details how she was disciplined by her department head because of other faculty's use of backstage communication channels:

> Heather C.: I was specifically yelled at by my department head about faculty in the department sending each other emails and text messages about department issues—not just me sending texts or emails they were included on, but ALL of the faculty doing that. My department head was angry that they were not included on those messages (I didn't know if they needed to be included or not). I was yelled at about emails and text messages I wasn't included on—it was just their general frustration with shadow networks. I was called names. I was just a punching bag for all that frustration. Immediately after that meeting (which I reported to the Ombuds office), I had a meeting with two junior faculty about graduate faculty work (at the time I was the graduate coordinator). In that meeting, the male faculty member suggested we bring a graduate council item to the faculty in our upcoming faculty meeting. I said they would need to include the department head on emails about that because I had just been yelled at for having these email conversations without the department head CCed on them before asking to be on the agenda. So, we created a fake email chain with the department chair CCed so it looked like we had the conversation in that email (not separately in another meeting). While it prevented me from getting yelled at again, it also made it look like my male colleague was the good team player because he sent the initial email (even though he was getting stage directions from me on how to prevent the department head from getting mad).

Heather C.'s story shows how a group engaged in shadow communication to counterstrike future reprimand by strategically creating communication in the legitimate network to meet the demands of the leadership. However, this example also identifies a potentially dangerous side effect—because the male colleague sent the email first, he was able to further integrate himself into the "inner circle" by following the unwritten rules of the department head. This ended up furthering the factional warfare experienced in Heather C.'s department. This story highlights important elements of shadow networks: Who speaks for the shadow network? What form of communication does it use when communicating with the legitimate network? If the ultimate goal of the shadow network is to work to identify alternative ways to deal with organizational issues, then eventually, they may need to communicate with leadership. The shadow networks scholarship does not discuss how shadow network members communicate their work with leadership or others not in the network. Additionally, the literature is silent on how shadow network members negotiate shadow and legitimate communication practices simultaneously. Heather C.'s story shows one outcome of those communication negotiations.

Implications

Our experiences with using Zoom to create and communicate within shadow networks underscore the power of shadow networks. Similar to previous research (e.g., Shaw 1997; Stacey 1996), our shadow networks were designed to accomplish tasks that were burdened by administrative bloat and red tape in the legitimate system during a crisis point. However, our narratives also shine a light on the ways in which shadow networks can be used to build coalitions and offer support to our colleagues. The communicative functions of shadow networks can be both task-oriented *and* relational. Ultimately, it is important for organizational members engaging in shadow network communication within the legitimate system to remember that this is a *strategic* communication experience, not random grapevine communication. Successful shadow networks need to identify the communication spaces, structures, and approaches that will help them achieve their goals.

Our experiences in and with shadow networks during the early stages of the Covid-19 pandemic helped to identify the dangers and insidiousness of factions and factional warfare. Behrens (2009) is the only author we could find who discusses the emergence of factions resulting from shadow networks and the impact of shadow networks in creating, challenging, or worsening existing factions. We identified counterstriking as a useful communication strategy shadow networks could use to preemptively challenge the legitimate system, but there are dangers to these actions. In the examples we provided, counterstriking was successful, but it also served to create wedges and warfare by making some organizational members appear legitimate and others not. Figuring out how to successfully engage in counterstriking while avoiding discipline and furthering faction warfare is necessary. Shadow networks, by definition, have the ability to protect and provide voice to those in marginalized and oppressed positions within organizations, but caution ought to be noted in that these networks can also reify and exacerbate extant power differentials.

Finally, all the authors engaged in shadow network communication, but not all the authors were ordinary organizational members. Two of the authors were in prominent leadership positions and had a direct hand in Covid-19 decision-making at their universities. The previous shadow network research would position them as part of the legitimate system, not as average organizational members. However, organizations are not as simplistic as "leadership" and "members." There are multiple layers of leadership in institutions, and organizational members can embody multiple roles. This is especially true in higher education, where individuals can simultaneously hold positions as faculty/ staff and mid-level leadership positions (as was our case). This raises questions about the structure of shadow networks. Who really is the "authority" they are challenging? Can workers in leadership positions (but who still have leaders/ authority higher than them) engage in shadow networks? We would argue that the complex nature of hierarchy, authority, and power in organizations means the answer is yes.

Reflections and Conclusions

At the time of this writing, we just passed the third anniversary of the initial Covid-19 lockdowns, and scholars are undoubtedly engaging in reflexive thinking about the forms, channels, and ramifications of habits and patterns of communication adopted during the pandemic—many of which continue now. Workplace communication, including in higher education, has undoubtedly shifted and with it, the ways in which shadow networks emerge and function have adapted (and, perhaps even grown). Shadow networks, from a critical perspective, help enable silenced voices, can create shared spaces and a sense of belonging, and provide pushback within hierarchical structures. Shadow networks, as intrinsically outside the box, have the ability to promote creative problem solving, which was desperately needed during the pandemic. We recognize that these networks cannot be inclusive by their nature and thereby are prone to faction warfare as those in the in-group may clearly demarcate communication parameters. Additionally, we seek to further problematize the notions of *leadership* and *management* and the approach of both in entering and/or disciplining the shadow network. The ways in which administration allows (or foolishly does not allow) the shadow network to examine and pose solutions to problems will be imperative, particularly in higher education, where significant financial and political constraints continue. The disciplining or the construction of communication roadblocks for shadow networks deprives the system of bottom-up decision-making, reinforces ivory tower and balcony administrative metaphors, and further entrenches division between workers and leadership. As we continue to come back together in physical and digital spaces, we must work to understand how shadow networks can help workers to communicate and challenge effectively to create workspaces, technologies, and communication structures that work for everyone.

References

Behrens, Sandra. 2009. "Shadow Systems: The Good, The Bad, and the Ugly." *Communications of the ACM* 52, no. 2: 124–9. https://doi.org/10.1145/1461928.1461960.

Burke, Lisa and Jessica Wise. 2003. "The Effective Care, Handling, and Pruning of the Office Grapevine." *Business Horizons* 46, no. 3: 71–6. https://doi.org/10.1016/S0007-6813(03)00031-4.

Crampton, Suzanne, John Hodge, and Jitendra Mishra. 1998. "The Informal Communication Network: Factors Influencing Grapevine Activity." *Public Personnel Management* 27, no. 4: 569–84. https://doi.org/10.1177/009102609802700410.

Ellingson, Laura. 2003. "Interdisciplinary Health Care Teamwork in the Clinic Backstage." *Journal of Applied Communication Research* 31, no. 2: 93–117. https://doi.org/10.1080/0090988032000064579.

Ellingson, Laura. 2004. *Communicating in the Clinic: Negotiating Frontstage and Backstage Teamwork*. New York: Hampton Press.

Foucault, Michel. 1995. *Discipline and Punish: The Birth of the Prison*. 2nd ed. New York: Vintage Books.

Fox, Jesse, and Bree McEwan. 2017. "Distinguishing Technologies for Social Interaction: The Perceived Social Affordances of Communication Channels Scale." *Communication Monographs* 84, no. 3: 298–318. https://doi.org/10.1080/03637751.2017.1332418.

Goffman, Erving. 1959. *The Presentation of Self in Everyday Life*, New York: Doubleday.

Kadushin, Charles, and Michael Brimm. 1990. *Why Networking Fails: Double Binds and the Limitations of Shadow Networks*. Paper presented at the 10th Sunbelt International Social Network Conference in San Diego, CA.

Kast, Giovanna. 2019. "Grapevine Communication in Communication Centers: The Needs and Effects." *Communication Center Journal* 5, no. 1: 181–3. https://libjournal.uncg.edu/ccj/article/view/1959/pdf.

Manley, Andrew and Shaun Williams. 2022. "We're Not Run on Numbers, We're People, We're Emotional People": Exploring the Experiences and Lived Consequences of Emerging Technologies, Organizational Surveillance, and Control Among Elite Professionals." *Organization* 29, no. 4: 692–713. https://doi.org/10.1177/1350508419890078.

Mascheroni, Giovanna and Jane Vincent. 2016. "Perpetual Contact as a Communicative
    Affordance: Opportunities, Constraints, and Emotions." *Mobile Media &
    Communication* 4, no. 3: 310–26. https://doi.org/10.1177/2050157916639347.

Monge, Peter, and Noshir Contractor. 2001. "Emergence of Communication Networks."
    In *The New Handbook of Organizational Communication: Advances in Theory,
    Research, and Methods*, edited by F. M. Jablin and L. L. Putnam, 440–502. Newbury
    Park: Sage.

Robinson, Katy and Patrick Thelen. 2018. "What Makes the Grapevine so Effective?
    An Employee Perspective on Employee-Organization Communication and Peer-
    To-Peer Communication." *Public Relations Journal* 12, no. 2: 1–20. https://prjournal
    .instituteforpr.org/wp-content/uploads/Robinson_Thelen_What-Makes-the
    -Grapevine-So-Effective.pdf.

Salem, Philip. 2021. "Transformative Organizational Communication Practices" In
    *Research Anthology on Digital Transformation, Organizational Change, and the
    Impact of Remote Work*, edited by Information Resources Management Association,
    931–51. Hershey: IGI Global.

Sarai, Noreen and Tatenda Gotora. 2021. "Effectiveness of Grapevine as a
    Communication Strategy in Tertiary Administration in the Dynamic World
    of Social Media: COVID-19 Pandemic." *International Journal of Research and
    Innovation in Social Science* 5, no. 2: 386–92. https://www.rsisinternational.org/
    journals/ijriss/Digital-Library/volume-5-issue-2/386–392.pdf.

Schouten, Alexander, Patti Valkenburg, and Jochen Peter. 2007. "Precursors and
    Underlying Processes of Adolescents" Online Self-Disclosure: Developing and
    Testing an 'Internet-Attribute Perception' Model." *Media Psychology* 10, no. 2:
    292–315. https://doi.org/10.1080/15213260701375686.

Sewell, Graham and James Barker. 2006. "Coercion Versus Care: Using Irony to Make
    Sense of Organizational Surveillance." *Academy of Management Review*, 31, no. 4:
    934–61. https://doi.org/10.5465/amr.2006.22527466.

Shaw, Patricia. 1997. "Intervening in the Shadow Systems of Organizations: Consulting
    from a Complexity Perspective." *Journal of Organizational Change Management*, 10,
    no. 3: 235–50. https://doi.org/10.1108/09534819710171095.

Stacey, Ralph. 1996. *Complexity and Creativity in Organizations*. Oakland: Berrett-
    Koehler.

Tracy, Sarah J. 2000. "Becoming a Character for Commerce: Emotion Labor, Self-
    Subordination, and Discursive Construction of Identity in a Total Institution."
    Management Communication Quarterly 14, no. 1: 90–128. https://doi.org/10.1080
    /02763890802665007.

Wittenberg-Lyles, Elaine, Ginnifer Cie'Gee, Debra Oliver, and George Demiris. 2009.
    "What Patients and Families Don't Hear: Backstage Communication in Hospice
    Interdisciplinary Team Meetings." *Journal of Housing for the Elderly* 23, no. 1–2:
    92–105. https://doi.org/10.1080/02763890802665007.

Unstable Connections

In this final section, we attempt to unpack the logic of "connection" embedded within Zoom and other videoconferencing tools—not only the degree to which a technosolutionism (Morozov 2013) based upon "universal access" underpins its ideology of efficient performance but also how it assumes as its user a neoliberal "enterprising subject" (Houghton 2019) who can deploy these networks to their own advantage. But as we note in our introduction to this volume, a crisis within this system (such as a global pandemic) merely provides an opportunity to reaffirm the resilience of the performativity principle that underlies the system's ongoing operation. Yet we can likewise read in these disruptions the performative acts of the network logic itself—a logic that calls into being through modulations in control a "new normal," declaring *this is fine*[1] while the world burns. Moments of stoppage within the ceaseless circulation of communicative capitalism may very well signal events in which marginalized groups find themselves pushed further to the margins, but these events likewise take place within the operation of technological and ideological apparatuses "upon which claims to universality are raised and defended" (Dean 2009, 22). Ironically, then, calls for greater access are likewise invitations into systems of heightened surveillance. Promises of egalitarian stability operate as an ideological veneer, drawing attention away from the ongoing precarity of these individuals now caught up in its apparatuses of control. As Dean (2009) notes, "The values heralded as central to democracy take material form in networked communications technologies. Ideals of access, inclusion, discussion, and participation come to be realized in and through expansions, intensifications, and interconnections of global telecommunications" and the disparities in access to power that these same networks support and promote (23).

At the fore are issues of privilege—access not just in the sense of "being online" but also access to the socioeconomic and cultural situatedness that positions "stable connection" as a default mode of being for the Zoom subject (if only one's voice would be heard and acknowledged by simply clicking the "unmute" button). While the move to daily life online during Covid-19 was a disorienting experience for many, "success" in orienting to the platform had less to do with technological savvy than the privilege for which digital literacy serves as a proxy measure: "the ability to move through the world without losing one's way" (Ahmed 2006, 139). That is not to say that supportive communities—from twelve-step recovery programs to queer dance parties—did not find a *place* to gather during quarantine; indeed, these emerged as the same precarious "togetherness" that Cochrane and Kenney discuss in their contribution to this collection (and which we expand on in our own contribution in this section). Here as well, we would point to the haphazard and tentative performative elements that allow these networks to form but also threaten their stable existence. The fault line that marks this troubled stability maps a familiar trajectory of privilege and power: to see these threats erupt, one need only look to the predominance of female teachers targeted for zoombombing (Elmer et al. 2021) and the sexist, racist, xenophobic, homophobic, and transphobic content of these attacks during a period in which mediated presence offered an alluring promise of solidarity and social support (Urbach 2020; Nakamura et al. 2021; Ali 2021).

The Covid-19 pandemic made visible for those with the privilege of "stable connection" as a default setting what has been a lived reality for many for a long time: those communities and individuals who are most in need of services and support are the ones to experience the greatest degree of digital precarity. As Kennedy, Holcombe-James and Mannell (2021) note in their discussion of digital inclusion and exclusion, material access to technology alone by no means ensures an individual's access to the "more subtle digital skills around the norms and conventions of video conferencing." Earvin Charles B. Cabalquinto and Evgeniya Pyatovskaya both take up this line of argument in this section through two specific case studies of different ethnic and racial groups in two different geographic settings. In "Locked In or Locked Out? Aging Migrants Enacting Autonomous and Dependent Copresence on Zoom during the Pandemic," Cabalquinto discusses the findings of a survey of older adults from culturally and linguistically diverse (CALD) backgrounds in Australia in 2020 and 2021. Cabalquinto notes the complex ways in which social, familial, and community networks were navigated with Zoom during the lockdowns and how the participants navigated being present with loved ones and community members over Zoom. In particular, these interview participants discussed the benefits of the technology and how it helped them to participate in community events and to connect with family members abroad. They also discussed, however, the challenges they faced, such as a lack of technological competency and limited support in navigating this new technology. Cabalquinto highlights the importance of understanding who benefits from Zoom as well as who *could*, if given the opportunity to do so.

Through a case study on a nonprofit organization in New York City that serves Asian immigrant (AI) and Asian American (AA) youth, Pyatovskaya's chapter, "'Zoom Saved All Our Lives': A Case of Nonprofit Resilient Organizing during the Covid-19 Pandemic," highlights the very real social, class, and technological inequities facing these youth and their families before and during the pandemic. Pyatovskaya's case study identifies how lack of access to technology was only part of the issue; knowledge and know-how of utilizing tech, often across age, culture, and language boundaries, made the use of Zoom challenging for the organization's programming. Much like Cabalquinto's chapter, Pyatovskaya highlights what she refers to as a "complex matrix of oppression" wherein previous barriers were reinforced and the gap in access to supports widened, particularly for immigrant, multigenerational families. "Resilience"—that celebrated trait of distributed networks and neoliberal subjects alike—serves as a necessary survival strategy for precarious groups, but it likewise marks and makes visible the ideological and technological apparatuses that marginalize these communities and necessitate resilience in the first place.

In our own contribution to this collection, "(Un)expected Errors," we explore more explicitly the instability of connections and connectivity that is embedded within the logic of maximum performance. We examine zoombombing in greater depth, but we also explore what happens when performativity tools of the enterprising digital subject are turned against these same apparatuses of control: such as when students embrace a cameras-off aesthetic or when the tools designed to facilitate working at home are deployed as anti-work tactics. We also, as best we can, attempt to answer some of our own lingering questions and process our own experiences over the past three years. Indeed, the whole of this volume was developed, edited, and written by both of us in different physical locations and time zones but evolved through regular Zoom meetings, piecing together not only the works in this volume but also our own professional and personal space for precarious togetherness. Ironically, it is the unstable connections that brought us together and, much like the other chapters in this section, led us to powerful adaptations in a messy world.

Thus, we end on a surprisingly high note: a hope for what-might-be. As Tania Lewis, Annette Markham, and Indigo Holcombe-James (2021) note in their discussion of the liminality of our experiences living through an era of Covid-19: rather than seeing the "end" of the pandemic as a *return to normal*, might we instead seek to "challenge the assumption that stability and certainty is what we now need as a global community"? They go on to ask: "[H]ow can we use the discomfort of liminality to imagine global futures that have radically transformative possibilities?" We find ourselves, on and off Zoom, in just such a precarious moment. With these insights in mind, we can hardly claim stable ground upon which to draw conclusions. Instead, we hope that the conversations we have engendered here will continue in pursuit of these potentials and openings—no matter how haphazard, unstable, or temporary they may seem.

Note

1 For discussion of the relevance of this meme to the life and times of individuals living under a logic of Zoom during "pandemic times," see Cochrane and Kenney, in this volume.

References

Ahmed, Sara. *Queer Phenomenology: Orientations, Objects, Others.* Durham: Duke University Press, 2006.

Ali, Kawsar. 2021. "Zoom-ing in on White Supremacy: Zoom-Bombing Anti-Racism Efforts." *M/C Journal* 24, no. 3. https://doi.org/10.5204/mcj.2786.

Dean, Jodi. 2009. *Democracy and Other Neoliberal Fantasies: Communicative Capitalism and Left Politics.* Durham: Duke University Press.

Elmer, Greg, Stephen J. Neville, Anthony Burton, and Sabrina Ward-Kimola. 2021. "Zoombombing During a Global Pandemic." *Social Media + Society* 7, no. 3. https://doi.org/10.1177/20563051211035356.

Houghton, Elizabeth. 2019. "Becoming a Neoliberal Subject." *Ephemera* 19, no. 3: 615–26.

Kennedy, Jenny, Indigo Holcombe-James, and Kate Mannell. 2021. "Access Denied: How Barriers to Participate on Zoom Impact on Research Opportunity." *M/C Journal* 24, no. 3. https://doi.org/10.5204/mcj.2785.

Lewis, Tania, Annette Markham, and Indigo Holcombe-James. 2021. "Embracing Liminality and 'Staying With the Trouble' On (and Off) Screen". *M/C Journal* 24, no. 3. https://doi.org/10.5204/mcj.2781.

Morozov, Evgeny. 2013. *To Save Everything, Click Here: The Folly of Technological Solutionism.* New York: PublicAffairs.

Nakamura, Lisa, Hanah Stiverson, and Kyle Lindsey. 2021. *Racist Zoombombing.* New York: Routledge.

Urbach, Martin. 2020. "Zoombombing is a Mirror that Forces Us to Look at Ourselves." *Medium*, April 4, 2020. https://medium.com/@martinurbach/zoombombing-is-a-mirror-that-forces-to-look-at-ourselves-f6d10cf6fa2e.

10

Locked In or Locked Out? Aging Migrants Enacting Autonomous and Dependent Copresence on Zoom during the Pandemic

Earvin Charles B. Cabalquinto, Monash University

Introduction

The onset of the global health pandemic has exposed the vital role of digital technologies, social media, and mobile applications in sustaining gatherings and interactions disrupted by perennial lockdowns, travel bans, and cross-border shutdowns. In the first instance, individuals dispersed across and within nations used and accessed digital technologies to maintain personal, familial, and professional connective needs (Hargittai 2022). However, other individuals faced multiple communicative constraints due to their limited digital access and low digital literacy (Hargittai 2022).

This chapter presents how fifteen older adults from culturally and linguistically diverse (CALD) backgrounds used Zoom during a series of lockdowns in 2020 and 2021 in Victoria, Australia. Zoom was one of the online channels utilized by this cohort of people to forge and maintain personal, familial, and social connections within and beyond Australia. Notably, older people from CALD backgrounds belong to the 1.2 million overseas-born Australians reported in the 2016 Australian Bureau of Statistics (ABS). They also represent over one-third (37 percent) of the older Australian population (Australian Institute of Health and Welfare 2023). Over the past several years, many studies have shown how aging CALD people in Australia rely on digital communication technologies and online channels to maintain ties with their family members and peers in a national and transnational context (Baldassar and Wilding 2019; Baldassar, Wilding, and Worrell 2020), even before the pandemic. However, studies have also highlighted that digital behaviors depend on the aging migrant's digital access, resources, capabilities, and support networks (Wilding et al. 2022; Baldassar et al. 2020). Nevertheless, scholarly works underline the interest and motivation of aging migrants to use modern communication technologies as part of their personal, familial, and social life. Within this context, this chapter illuminates the ways aging migrants used Zoom to maintain copresence with their distant networks. The discussion is based on the collected data drawn from deploying remote interviews using both Zoom and phone calls: fifteen aging participants interviewed in 2020, ten of whom engaged in follow-up interviews in 2021.

I deploy a socio-material approach to examine the role of Zoom in facilitating mediated copresence between the aging migrants and their distant local and transnational networks. This perspective highlights how the interlinking of technological devices, social relations, and environments shapes the way people experience personal and social connections (Mol 2008; Pols 2012). An example of this perspective's applicability to understanding digital practices during the pandemic is the work of Watson, Lupton, and Michael (2020). Using a socio-material approach, they examined how Zoom allowed scattered Australian family members to remain connected with each other and maintain familial connections. Notably, within the growing body of literature on digital media and migration, a videoconferencing tool is key in enabling copresence at a distance (Cabalquinto 2022a; Francisco 2015; Longhurst 2013b; Madianou and Miller 2012). Through a socio-material approach, Ahlin (2020) shows how the quality of copresence care is shaped by the use of videoconferencing among transnational families. Importantly, scholars have also reiterated how the quality and outcomes of videoconferencing depend on people's access, resources, capabilities, and environments (Longhurst 2013b; Madianou and Miller 2012). In the case of older adults, autonomous and independent digital media use depends on one's access, competencies, and support networks (Fernández-Ardèvol 2020; Hänninen et al. 2021).

Drawing on these studies, the chapter examines the differing practices and capacities of aging migrants in using Zoom to enact mediated copresence. To articulate the findings of the study, I use the term "(In)Dependent Mediated Co-Presence" to show how aging migrants' differing digital access, competencies, and accessibility of support networks impact ways of being copresent with distant family members and peers. I contend that this articulation sets a vantage point to rethink how digital exclusion emerges, especially when mediated civic (Helsper 2021) and cultural participation (Ragnedda and Muschert 2018) among older migrants' access and capabilities are heavily reliant on their social and support circle.

Videoconferencing Copresence

In an increasingly digital and global society, modern communication technologies and online platforms are fundamental in shaping people's personal, familial, and social lives. This holds true especially among migrants and their distant family members and networks, who rely heavily on digital technologies to forge and maintain ties within and beyond borders (Baldassar, Baldock, and Wilding 2007; Cabalquinto 2022a; Madianou and Miller 2012). Indeed, communication technologies allow individuals to stay connected and experience solidarity and belongingness (Diminescu 2008). And during the pandemic, technological apparatuses functioned as lifelines, enabling migrants, their kin, and their networks to endure the uncertainty and precarity of physical separation created by the imposition of lockdowns, travel bans, and border closures.

This chapter investigates how aging CALD people utilized Zoom, one of the preferred videoconferencing tools during the lockdown. As the global health pandemic put many people's mobility on hold, studies have shown the widespread uptake of Zoom among diverse cohorts of people and personal and professional arrangements (Hargittai 2022; Kennedy et al. 2021; Nunes and Ozog 2021). Videoconferencing allowed physically separated people to restage a diverse range of activities, such as enabling family reunions online (Kaplan 2020) and weddings (Schiffer 2020), as well as sustaining learning and work commitments (Hargittai 2022). However, to date we know little about how older migrants used Zoom as part of their everyday lives during the pandemic. This gap also presents an opportunity to further interrogate how digital exclusion is experienced and navigated by vulnerable communities (Hargittai 2022) such as older migrants, considering their socio-demographic backgrounds, living conditions, and differing digital environments and capabilities (Helsper 2021).

I approach Zoom as a space that facilitates mediated copresence among geographically separate individuals. In the first instance, the networked and visual affordances of Zoom enable physically separated individuals to convene and converse at a distance. Notably, Zoom's key affordance builds on a key infrastructure—the webcam. As noted by Madianou and Miller (2012), "[the] webcam is not only interactive (which is linked to its temporality), but also provides strong visual cues (facial expressions, tone of voice, pauses, gestures) that contextualize communication" (134). In this vein, individuals can replicate face-to-face interactions, collapsing differences in time and space through which individuals are situated. At the same time, the screen of the mobile device also functions as a place in itself, entangled with data and infrastructures that facilitate ways of being present with distant others (Richardson and Wilken 2012).

Within the growing study on the intersections of migration and digital media, research studies show how videoconferencing tools have been used for "streaming" copresence and allow the maintenance of familial ties (Cabalquinto 2022a; Francisco 2015; Longhurst 2013a; Madianou and Miller 2012). In the intersections of digital media, migration, and aging, research shows the ways videoconferencing has been accessed by aging parents to assess the health and well-being of their distant children and then provide advice as a form of care (Ahlin 2020). Notably, the visual affordance of the webcam is relied upon to achieve the performance of familial roles and obligations (Miller and Sinanan 2014), such as interacting to deliver care and support among dispersed family members (Longhurst 2013b; Madianou and Miller 2012; Cabalquinto 2022a; Francisco 2015). Notably, videoconferencing has allowed intimate one-to-one or multiple interactions at a distance (Miller and Sinanan 2014). This has been evident in the impact of using Zoom during the pandemic on personal, educational, and professional arrangements (Nunes and Ozog 2021).

This chapter applies a socio-material approach to enact mediated copresence. In this perspective, scholars have underscored the use of digital technologies shaping the way people interact, socialize, and connect (Mol et al. 2015; Pols 2012). For instance, the networked affordances of digital technologies mediate interactions and facilitate proximity among patients and their caregivers in a hospital or home care arrangement (Pols 2012). In a formal setting, dispersed individuals use a range of technologies to enact one's presence as well as deliver care and intimacy (Pols 2012). Meanwhile, in an informal setting such as the household, Watson, Lupton, and Michael (2020) have shown how the use of Zoom among Australian families during the pandemic allowed dispersed family members to gather, interact on the screen, and deliver familial support through synchronous communication.

It is worth noting that the quality and outcomes of videoconferencing are also dependent on a range of social and technological factors. This point exposes how the intertwining of social and digital inequalities manifests in using Zoom (Hargittai 2022; Kennedy, Holcombe-James, and Mannell

2021). In the first instance, the operation of such a medium depends on an individual's infrastructural and digital access (Cabalquinto 2022a; Madianou and Miller 2012). For example, studies have shown that the unstable internet connection between the host and home countries of migrants often undermines the quality of videoconferencing (Cabalquinto 2022a). In this case, distant family members opt to switch off their camera to get a better and more clear connection (Cabalquinto 2022a). During the pandemic, individuals who did not have a stable internet connection struggled with communicating at a distance via Zoom (Nunes and Ozog 2021), and some people even had to leave their houses to access free and public Wi-Fi service (Hargittai 2022). Additionally, individual users with mobile-only data struggled with videoconferencing as such platforms often require stable and fixed broadband connectivity to work effectively (Kennedy et al. 2021). In some cases, an individual's lack of skills in videoconferencing created frustration and eventual nonuse, as experienced by aging parents of migrants (Cabalquinto 2022a). Indeed, a videoconferencing tool might offer "telepresence" (Richardson and Wilken 2012) or "attainment" of presencing despite the distance (Miller and Sinanan 2014), positioning it as a "sunny day technology" given the positive outcomes it brings to interlocutors (Wilding 2006). Yet, disparities in accessing infrastructures and resources as well as capabilities of individual users limit and undermine the use of Zoom (Hargittai 2022; Kennedy et al. 2021; Nunes and Ozog 2021). Indeed, this point illuminates who gets to participate or not in an increasingly digital society (van Dijck 2020) as shaped by an individual's background as well as their social and digital environments (Helsper 2021, Ragnedda and Muschert 2018). And in this study, I demonstrate that the (in)dependent mediated copresence performed and experienced by aging migrants in using Zoom is deeply informed by disparities in digital access, competencies, and accessible support networks.

Methodological Approach

This chapter is based on data drawn from conducting remote interviews—via Zoom or phone—among fifteen older people from CALD backgrounds during 2020 and 2021.[1] I conducted fifteen interviews in 2020: nine over the phone and six on Zoom. Ten of the fifteen participants were engaged in follow-up interviews in 2021. The participants had Australian citizenship status. The ages ranged from sixty to ninety-six years old ($N = 71$). There were nine females and six males. Most participants had Asian backgrounds, indicating a growing number of older migrants with Asian backgrounds in Australia (Wilson et al. 2020). Based on the 2016 ABS, there were 1.2 million overseas-born Australians, representing over one-third (37 percent) of the older Australian population (Australian Institute of Health and Welfare 2021). It is projected that by 2056, the combined number of 65+ Asian-born individuals will contribute 19.1 percent (up from 6 percent in 2016) to Australia's total older population (Wilson et al. 2020).

The data collection focused on mapping the digital practices of the participants during the lockdown in Victoria, Australia. Victoria had the longest lockdown in the world, with 260 days during 2020 and 2021. In 2020, people were only allowed to leave their houses to buy essential goods, exercise for a limited time, study, work when impossible to do at home, and buy essential grocery items. In 2021, receiving a vaccination was added to the four provisions to leave one's house. In unraveling the living conditions of the participants, the interviews also revealed the motivations and barriers behind the participants' digital media use. Notably, Zoom was one of the many online channels utilized by the participants to stay connected with their family members and peers in Australia and overseas. The collected data was analyzed through a thematic analysis (Creswell 2013), paying attention to the similarities and differences in utilizing Zoom. Through such an approach, the findings revealed the diverse motivations, practices, and negotiations of the participants in using Zoom, to be discussed later.

The study utilized pseudonyms to protect the privacy of the participants. I also used pseudonyms to deidentify the names and places mentioned by the participants during the interview. I selected the quotes included in this chapter, with most quotes edited for clarity. The project obtained ethics clearance from the Deakin University Human Research Ethics Committee, reference number HAE-20–106.

Life on Zoom during a Lockdown

> Now that everybody cannot go out, we have Zoom. Once in a couple of weeks' time, and I can see all my family and I can talk to them [*referring to using Zoom*]. And it makes me so happy.

This quote is from Anna, an 85-year-old Malaysian-Australian. She moved to Australia in 1977 with her husband and three children. The move was motivated by their goal to provide a better education for their children. During the interview, Anna shared that she had a leg operation when she was seventy-three years old, which limited her mobility when moving out and about. She currently lives with her daughter and her daughter's family. Living with them has provided her with emotional and physical support. Notably, her daughter and son-in-law taught her to use her mobile device and a range of online channels. Digital devices allow her to connect to her peers in Australia as well as relatives in Malaysia and Singapore. During the pandemic, she accessed Facebook, WhatsApp, YouTube, and Zoom to address her personal, familial, and social needs. As noted in her quote here, she relied on Zoom, as one of her online channels, to remain copresent with her family members and networks in Australia and overseas.

The case of Anna illustrates the integral role of Zoom in enabling aging migrants to stay connected with their family members while coping with the uncertainty and immobility of the global health pandemic. As for most participants, Zoom, with its visual and networked affordances, has also been a "blessing" (Horst 2006) for achieving mediated copresence. Notably, even before the pandemic, the participants already utilized digital technologies and social media channels to activate mediated copresence among their dispersed peers and family members. Particularly, they used messaging applications for videoconferencing within their networks. However, during the pandemic, several of the participants began accessing and using Zoom under the sway of family members and peers who used Zoom.

For instance, Ella, the 65-year-old Taiwanese-Australian, started using Zoom with the encouragement of her siblings who are based in Australia and overseas. Before using Zoom, she and her family used a messaging application to videoconference. But her family members decided to use Zoom given its growing popularity. As she shared, "My elder sister said 'Yeah, why don't we get on Zoom?', and that's how it came about." Meanwhile, Mohammed, a 77-year-old Syrian-Australian, used Zoom based on the influence of his social circle. Notably, his understanding of videoconferencing was informed by using messaging applications to video call with his distant children and grandchildren overseas. During the interview, he revealed that he started using Zoom to connect with his peers in Australia. He said, "I am using the Zoom program because I am now dealing with another group. I am using this Zoom." Indeed, these examples show the relational dimension (Madianou and Miller 2012) of technology use in enacting mediated copresence.

However, mediated copresence enabled by Zoom is shaped by differing levels of access and competency, as well as differences in support networks. In the following sections, I will discuss subjects' experiences with and performances of (in)dependent mediated copresence.

Combating Isolation

In Victoria, stay-at-home orders prohibited individuals from house visits. It was only with the government's implementation of the "single-person bubble" policy in 2020 that individuals living alone could create a "bubble" with one other person. But for most part of the lockdowns, the participants stayed at home and accessed Zoom to connect with their local family members and peers. The participants relied heavily on their informal networks in setting up and eventually using Zoom. I refer to this as dependent mediated copresence, or ways of being copresent using Zoom through the assistance of an informal network. This concept builds on studies that have shown that older adults (Dolničar et al. 2018; Hänninen et al. 2021) and older migrants rely on their informal and support networks to set up and use technologies (Selwyn et al. 2016, Worrell 2021). Broadly speaking, informal networks assist the digital needs of older family members, such as purchasing of devices, installing and configuring software, and instructing on the use of technologies (Hänninen et al. 2021; Tolmie et al. 2007). For the participants of the study, family members were relied upon in setting up Zoom.

A case in point is Rosa, a 79-year-old Macedonian-Australian who utilized Zoom to stay connected with her family members. Rosa moved to Australia in 1967 and worked in a clothing factory. Her husband passed away in 2019 after battling dementia for seven long years. Dealing with arthritis in her back had limited her movements even before the pandemic, and she would normally only walk in her front yard. Pre-pandemic, her children and grandchildren would visit her on a regular basis. Unfortunately, physical visits were curtailed because of stay-at-home orders. To manage the physical separation, Rosa shared her experiences by having lunch parties via Zoom. During the interview, the granddaughter was at Rosa's place. This visit was allowed and acceptable, labeled as a caring duty and a "one-person bubble." I asked Rosa if she was using Zoom. The granddaughter assisted her by translating her answer to English: "She doesn't know how to use Zoom." Rosa confirmed: "No, I can't use Zoom." I then asked,

"So how do you join the party?" The granddaughter explained: "I come here and help her with it [Zoom], and stay for the chat with the rest of the family. There needs to be usually just one other person here. We'll go to the supermarket and get her things that she needs, and then come and set it up. So that's basically how it works." I asked Rosa how she feels about it. She responded that she's happy to be with her family using Zoom via the computer.

Another case is Lena, a 68-year-old Indian-Australian, who relied on her son to use Zoom and stage a cultural and community-based ritual, *Badjam*:

> We have a *Badjam*. So, we pray to God. Every month we have somebody in our house. Thirteen families come together and sing songs and pray to God and all eat together. And everyone brings one dish. And then we'll share everything, eat everything, and spend time, and we'll come back, and the night is spent. But at this time, we can't go out. My son makes the arrangements only for the prayer. We have families coming into the Zoom call. We finish everything.

In these examples, the participants relied on their family members to access and use Zoom and participate in mediated copresence. Rosa had to rely on her granddaughter to participate in the lunch party on Zoom and combat isolation. Meanwhile, Lena also had to depend on his son to set up Zoom and organize a cultural activity, *Badjam*. Indeed, the informal support networks in the households provided technological assistance among less technologically competent family members (Hänninen et al. 2021).

The participants used Zoom to connect to their transnational family members. This was commonplace given the travel ban imposed by countries around the world to fight the spread of Covid-19. Notably, there were participants who were adept at accessing Zoom. This ability provided them with independent mediated copresence. An example of this is Ella. During the pandemic, she was living alone. Her sister couldn't visit her because of the lockdown. From time to time, her nephew, a doctor, checked on her via phone call. She had family members in the United States, Taiwan, and across Australia, but given the travel ban, overseas travel was impossible. She was one of the participants who was living alone and had technological competency in accessing and using an online platform. She accessed and used Zoom by downloading it to her smartphone.

During the interview, Ella shared her experiences of using Zoom to connect with her overseas family members. She began her statement by saying: "At this moment during the COVID we actually have a home Zoom meeting, a family Zoom meeting, so we use Zoom." I asked her to elaborate on the arrangement on Zoom. She explained, "We meet up once every two weeks and because everybody's living at a different spot in the world, so we just make it just like everybody can dial in and meet each other that way." For Ella, using Zoom was a positive experience: "It's just a chit chat. With a few of us around there's always finding things to chit chat [. . .] It's been positive." In this case, echoing studies on the use of videoconferencing in transnational familial communication (Ahlin 2020; Alinejad 2019; Cabalquinto 2022a;), Ella and her kin across the world remained copresent with each other by using Zoom.

Fulfilling Duty in the Community

The participants did not only use Zoom to combat isolation. They also accessed it to fulfill duties in their communities. Among the fifteen participants, four of the participants were leaders of organizations for older migrants. One participant was heavily involved in volunteering by supporting villages in their home country. These participants used Zoom to establish copresence with their community members and deliver support. Furthermore, these individuals were relatively competent in using Zoom for communication, therefore affording them independent mediated copresence.

For example, I spoke to Maly, a sixty-year-old Cambodian-Australian and a leader and member of the organization for older Cambodian-Australians. Maly moved to Australia with her husband in 1995. She met her husband in Cambodia. She was a university student and her husband was working for UNICEF. Maly and her husband have two daughters; one lives with them, while the other daughter is based in Sydney. To date, Maly is a proactive leader of an organization for older Cambodians. Pre-pandemic, she normally organized cultural gatherings and coordinated meetings with other members of the organization. She still helps the organization generate funding as well as manage events for the older Cambodian-Australians. To accomplish these tasks pre-pandemic, she would normally take public transport to work with the organization. She did not mind this; as she said: "I feel it's important for me to work for the Cambodian Association, especially for my age. . . . If I can do any support through the community I would like to do it." However, during the lockdowns, she utilized Zoom to facilitate activities and engage older Cambodian-Australians. She explained:

Yesterday because we meet on Zoom. It's amazing to see sixty two participants. It's so easy when you meet on Zoom. Everyone can just join in. One woman even sitting in the car there. She turned on her computer having a meeting. I find it also can be very positive for me. I can join any meeting. It's no problem even I'm at home. But before I have to travel an hour drive, half an hour. I don't need to.

For some participants, videoconferencing enabled connections to immediate communities in their homeland. This was the experience of Edwin, the 66-year-old Indonesian-Australian. Edwin obtained a PhD in computer science at a university in New Zealand. He met his wife, a New Zealander with an English background, during that period. After finishing his studies, they moved to and lived in Indonesia for quite some time. In 1988, they moved to Canberra and lived there for twenty-eight years. In 2014, when their children moved to Melbourne, New York, and London, they felt the effects of an empty nest. They sold their house and went traveling for a few years in Australia, but they decided to settle in Melbourne when their daughter had her first child. They currently reside in Victoria. Edwin keeps busy through his active participation as a volunteer in the Diaspora Network, composed of Indonesians across the world. Prior to this, he volunteered under the Australian Volunteers for International Development led by the Department of Foreign Affairs and Trade (DFAT). Through that organization, he served on a nine-month volunteering project, teaching people in communities to set up and use a village information system.

To date, as an active member of the Diaspora Network, he has been helping a village in Flores, one of the islands east of Bali. He explained, "So we helped them, and they are using that information system as well that we provide but only in a very basic way." Edwin's passion in thinking about ways to help Indonesians is deeply linked to his Indonesian identity and his experiences growing up in an Indonesian village: "I lived in a village when I was a very young boy. I still have fond memories of what an idealized village is like (laughs)." The goal of his community support is, as per Edwin, "to help this village, which is at a low level, to become more—well in Indonesia, there's a lot of terms called 'smart village.' Whatever that is, that could mean anything, but the idea is to help that village to become whatever they want to be." To achieve this, Edwin explained, "We introduced them to Zoom. So, you can imagine this village with limited internet, so they engaged with us through electronic means. So, they got WhatsApp and they talk to us on Zoom." This work has also been complemented by Edwin's work with his team on helping other villages in Indonesia by providing them with information systems as well as assisting with installation and teaching how to use the system. He's also part of a Facebook group that assists 7,000 members and 10,000 villages in Indonesia, providing them information on using OpenSid, an open-source application that he and his team created. They also utilized videoconferencing: "Initially I was using Skype. We had been using Skype for some time. Because I work in the IT industry so way back we already started using some videoconferencing early on. I was then using WebEx, such a long way back. Then [I] used Skype. Then people started introducing [me] to Zoom." Indeed, in Edwin's case, Zoom has become a tool to provide support to communities and villages in Indonesia. In his words: "I'm doing something useful."

By examining the case of Maly and Edwin, we can see the role of Zoom in facilitating copresence among community members in Australia and overseas. It also shows the appropriation of Zoom for the purposes of enriching one's cultural identity and cultural connectedness (Baldassar et al. 2020). Importantly, their Zoom access and competencies also afforded them with independent mediated copresence without relying on their informal networks to set up and use Zoom.

Navigating Interrupted Copresence on Zoom

In the earlier sections, we have seen the positive outcomes of using Zoom among the participants. Copresence has been shaped by the entanglement of digital devices, online platforms, and social relations (Mol 2008; Pols 2012). Importantly, Zoom access, technological competencies, and support from informal networks inform independent and dependent mediated copresence. However, as to be discussed in this section, the participants also experienced issues while using Zoom. Their support networks helped along the way.

During one interview, Lena, whom we met earlier, spoke about her challenges with accessing Zoom to attend a singing session with a music group. Prior to the pandemic, Lena's son would tag along with her to watch an Indian concert. She enjoyed the concert because she admitted that she's fond of music, particularly learning a song and singing. However, the pandemic stopped the concerts. During the lockdown, Lena's son encouraged her to join a music group. He knew the music teachers. Lena also knew some of the members. She had encounters with them at a time when she went out with her friends to watch their concerts. Eventually, Lena found herself as a member of the music group.

As she explained, "In the lockdown, we can't go there [*referring to a class*] to learn music. They teach online and we have nearly 30 people in the group. We all sing a new song, and we also have learning music." Lena's son assisted in setting up Zoom so that Lena could attend the sessions. This allowed her to participate in singing traditional Carnatic music. Importantly, she valued the group as a space for belonging, where members exchanged relevant information. As she noted, "We're not only learning music, but we get updated with the current situation. They called me before and they talked about the vaccination. . . . Through the music group, other than music, they have built important current news." In this vein, Zoom provided multiple opportunities for a participant like Lena, such as sociality, leisure, and staying updated about Covid-19. However, Lena sometimes struggled in accessing Zoom. To address this, she sought the help of her son, who had been assisting her in her tech needs, "If there's any problem in the group, I will call my son normally. But my husband is near to me, so I ask him too. Any problem, Zoom is not loading, or I have to put the ID number, he will help."

It was also a family member who assisted Mohammed, whom we met earlier, to use Zoom. As mentioned earlier, Mohammed was using Zoom to connect with his peers. He also opted to be interviewed via Zoom. However, he also revealed in the interview that he relied on his son to set up Zoom, enabling the Zoom interview. The son would normally visit Mohammed and his wife to check on them and also assist in their needs. Installing Zoom was one of the modes of assistance. He said, "He installed Zoom in my computer and I only enter my ID number and this password, that's all." He also added, "With my age it's very difficult for me to follow the new program, the new devices. When I need any help I ask my son: 'Come, here is the problem.'" However, during times in which his son was not around, he received assistance from his wife. He explained:

> My wife is taking English courses in a university. She is taking an English course on Zoom. She installed Zoom and she's using it on a daily basis. Now she gives me some help when I need to use Zoom. Yesterday we had a meeting on Zoom. She said, "If you touch this button, it means that you agree" [. . .] I don't know this thing. My wife helps me to know what is the function of each button.

In these examples, we see how informal networks, such as family members, assist with the tech issues experienced by older migrants (Selwyn et al. 2016; Worrell 2021). However, informal networks are not only family members. Studies have shown that an informal support network can also be those within the community of an older person (Yu et al. 2023). For instance, Rahul, a 63-year-old Sri Lankan-Australian, used Zoom to deliver online teaching. Rahul was one of the three participants who were still working. He has a PhD in mathematics. He moved to Australia in 1970, and his wife and children eventually followed him. He had been teaching in a university for a long time. During the lockdown, he utilized Zoom for teaching. He said, "We were studying, we were still struggling, all teachers. But slowly trying to learn, and other younger people who are smarter, and they are more equipped with this knowledge, so they help." Indeed, Rahul's experiences showed how the support of his students assisted him while using Zoom for teaching.

Who's Locked Out on Zoom?

This chapter demonstrates how Zoom has allowed aging migrants following stay-at-home orders to maintain mediated copresence with their kin and peers, locally and transnationally. In thinking about mediated copresence through a socio-material approach, I used the term "(in)dependent mediated copresence" to encapsulate the autonomous and dependent use of Zoom by aging migrants as informed by disparities in access, competencies, and accessibility of support networks. It is worth noting that "dependent" Zoom use was a commonplace among the participants, receiving assistance from their informal networks. I argue that this scenario provides a critical vantage point to interrogate digital inclusion. In order to advocate for a digitally inclusive society, there is a need to consider the differing levels of technological access and competency, as well as differential access to support networks, of certain individuals and groups such as aging migrants, who often rely on others to access and use digital technologies effectively.

It is worth pointing out that the study has its limitations. First, the interviews were conducted remotely via Zoom and phone calls (Cabalquinto and Ahlin 2023). This arrangement therefore excluded people who were not tech savvy. Taking this into account, the study does not offer a generalized perspective on aging migrants' use of Zoom. However, it provides insights on the practices of this cohort of people that future studies can further interrogate. Second, most participants were mostly educated and were working in diverse professions prior to retirement. These determinants influenced digital access and capabilities. As such, future research can look at how aging migrants from low-income households navigate and use a videoconferencing tool in forging and maintaining personal, familial, and social ties. This proposition will further expose how the entanglements of structural and digital inequalities (Helsper 2021) manifest in everyday digital practices. A point of inquiry is how the use of a videoconferencing tool such as Zoom can be constrained by having mobile-only data or not having the financial capital to avail of monthly, fixed broadband connectivity (Kennedy et al. 2021). Fourth, the study has unpacked the relational use of Zoom as shaped by one's access, resources, and literacy (Madianou and Miller 2012), as well as support networks. Future research can also engage with the networks of aging migrants and how their practices contribute to the (dis)empowering dimensions of digital media use among aging migrants. Last, further studies can unpack the digital experiences of aging migrants who have small, shrinking, or absent support networks and how this arrangement impacts digital exclusion.

In summary, this chapter has illuminated the crucial role of Zoom in enabling aging migrants to stay copresent with their distant family members and peers during the lockdowns. However, it also exposes the dependence of this cohort of people for access to and use of a videoconferencing tool such as Zoom. These outcomes show the paradox of digital technology use shaped by unequal social and technological factors (Cabalquinto 2022b; Leurs 2014). Indeed, by zooming into the autonomous and shared digital practices of aging migrants, the study sheds light on a critical conversation regarding the meanings, practices, and outcomes of digital exclusion during pandemic times, and beyond.

Note

1   For a more complete discussion of methodology, and the impact of remote data collection on fieldwork, see Cabalquinto and Ahlin (2023).

References

Ahlin, Tanja. 2020. "Frequent Callers: 'Good Care' with ICTs in Indian Transnational Families." *Medical Anthropology* 39, no. 1: 69–82. https://doi.org/10.1080/01459740.2018.1532424.

Alinejad, Donya. 2019. "Careful Co-presence: The Transnational Mediation of Emotional Intimacy." *Social Media + Society* 5, no. 2: 1–11. https://doi.org/10.1177/2056305119854222.

Australian Institute of Health and Welfare. 2023. *Older Australians - Culturally and Linguistically Diverse Older People*. https://www.aihw.gov.au/reports/older-people/older-australia-at-a-glance/contents/demographics-of-older-australians/culturally-linguistically-diverse-people.

Baldassar, Loretta, Cora Baldock, and R. Wilding. 2007. *Families Caring Across Borders: Migration, Ageing and Transnational Caregiving*. Hampshire: Palgrave Macmillan.

Baldassar, Loretta and Raelene Wilding. 2019. "Migration, Aging, and Digital Kinning: The Role of Distant Care Support Networks in Experiences of Aging Well." *The Gerontological Society of America* 60, no. 2: 1–9. https://doi.org/10.1093/geront/gnz156.

Baldassar, Loretta, Raelene Wilding, and Shane Worrell. 2020. "Elderly Migrants, Digital Kinning and Digital Home Making Across Time and Distance." In *Ways of Home Making in Care for Later Life*, edited by Bernike Pasveer, Oddgeir Synnes and Ingunn Moser, 41–62. Singapore: Palgrave Macmillan.

Cabalquinto, Earvin Charles. 2022a. *(Im)mobile Homes: Family Life at a Distance in the Age of Mobile Media*. New York: Oxford University Press.

Cabalquinto, Earvin Charles. 2022b. "The Paradox of a Mobile Society: Situating Cultural Studies in the Global South Context." *International Journal of Cultural Studies* 26, no. 1: 22–33. https://doi.org/10.1177/13678779221134308.

Cabalquinto, Earvin Charles and Tanja Ahlin. 2023. "Researching (Im)mobile Lives During a Lockdown: Reconceptualizing Remote Interviews as Field Events." *International Journal of Cultural Studies*. https://doi.org/10.1177/13678779231157428.

Creswell, John W. 2013. *Qualitative Inquiry and Research Design: Choosing among Five Approaches*. Edited by John W. Creswell. 3rd ed. Thousand Oaks: Sage Publications.

Diminescu, Dana. 2008. "The Connected Migrant: An Epistemological Manifesto." *Social Science Information* 47, no. 4: 565–79. https://doi.org/10.1177/0539018408096447.

Dolničar, Vesna, Darja Grošelj, Maša Filipovič Hrast, Vasja Vehovar, and Andraž Petrovčič. 2018. "The Role of Social Support Networks in Proxy Internet Use from the Intergenerational Solidarity Perspective." *Telematics and Informatics* 35, no. 2: 305–17. https://doi.org/10.1016/j.tele.2017.12.005.

Fernández-Ardèvol, Mireia. 2020. "Older People Go Mobile." In *The Oxford Handbook of Mobile Communication and Society*, edited by Rich Ling, Leopoldina Fortunati, Gerard Goggin, Sun Sun Lim and Yuling Li, 187–99. Canada: Oxford University Press.

Francisco, Valerie. 2015. "'The Internet Is Magic': Technology, Intimacy and Transnational Families." *Critical Sociology* 41, no. 1: 173–90. https://doi.org/10.1177/0896920513484602.

Hänninen, Riitta, Laura Pajula, Viivi Korpela, and Sakari Taipale. 2021. "Individual and Shared Digital Repertoires – Older Adults Managing Digital Services." *Information, Communication & Society* 26, no. 3: 1–16. https://doi.org/10.1080/1369118X.2021 .1954976.

Hargittai, Eszter. 2022. *Connected in Isolation: Digital Privilege in Unsettled Times.* Cambridge, MA: Massachusetts Institute of Technology.

Helsper, Ellen. 2021. *The Digital Disconnect: The Social Causes and Consequences of Digital Inequalities.* London: SAGE.

Horst, Heather. 2006. "The Blessings and Burdens of Communication: Cell Phones in Jamaican Transnational Social Fields." *Global Networks* 6, no. 2: 143–59. https://doi .org/10.1111/j.1471–0374.2006.00138.x.

Kaplan, Melanie. 2020. "When Friends and Family can't Travel, Zoom Reunions Provide an Emotional (and virtual) Connection." *Washington Post*, April 16, 2020. Available online: https://www.washingtonpost.com/lifestyle/travel/when-friends -and-family-cant-travel-zoom-reunions-provide-an-emotional-and-virtual -connection/2020/04/16/bc3a22f4-7da6-11ea-9040-68981f488eed_story.html.

Kennedy, Jenny, Indigo Holcombe-James, and Kate Mannell. 2021. "Access Denied: How Barriers to Participate on Zoom Impact on Research Opportunity." *M/C Journal* 24, no. 3. https://doi.org/10.5204/mcj.2785.

Leurs, Koen. 2014. "The Politics of Transnational Affective Capital: Digital Connectivity among Young Somalis Stranded in Ethiopia." *Crossings: Journal of Migration & Culture* 5, no. 1: 87–104. https://doi.org/10.1386/cjmc.5.1.87_1.

Longhurst, Robyn. 2013a. "Stretching Mothering: Gender, Space and Information Communication Technologies." *Hagar* 11, no. 1: 121–38, 186.

Longhurst, Robyn. 2013b. "Using Skype to Mother: Bodies, Emotions, Visuality, and Screens." *Environment and Planning D: Society and Space* 31, no. 4: 664–79. https:// doi.org/10.1068/d20111.

Madianou, Mirca and Daniel Miller. 2012. *Migration and New Media: Transnational Families and Polymedia.* Abingdon, Oxon: Routledge.

Miller, Daniel and Jolynna Sinanan. 2014. *Webcam.* Cambridge: Polity Press.

Mol, Annemarie. 2008. *The Logic of Care: Health and the Problem of Patient Choice.* New York: Routledge.

Mol, Annemarie, Ingunn Moser, and Jeannette Pols. 2015. *Care in Practice: On Tinkering in Clinics, Homes and Farms.* Bielefeld: transcript Verlag.

Nunes, Mark and Cassandra Ozog. 2021. "Your (Internet) Connection Is Unstable." *M/C Journal* 24, no. 3. https://doi.org/10.5204/mcj.2813.

Pols, Jeannette. 2012. *Care at a Distance: On the Closeness of Technology, Care & Welfare.* Amsterdam University Press.

Ragnedda, Massimo and Glenn W. Muschert. 2018. *Theorizing Digital Divides, Routledge Advances in Sociology.* New York: Routledge.

Richardson, Ingrid, and Rowan Wilken. 2012. "Parerga of the Third Screen: Mobile Media, Place and Presence." In *Mobile Technology and Place*, edited by Gerard Goggin and Rowan Wilken, 198–212. New York: Routledge.

Schiffer, Zoe. 2020. "Saying 'I do' over Zoom." *The Verge*, April 1, 2020. https://www.theverge.com/2020/4/1/21202466/zoom-wedding-coronavirus-covid-19-social-distance.

Selwyn, Neil, Nicola Johnson, Selena Nemorin, and Elizabeth Knight. 2016. "Going Online on Behalf of Others: An Investigation of 'Proxy' Internet Consumers." *Australian Communications Consumer Action Network*, April 2016. Available online: https://accan.org.au/files/Grants/ACCAN_Monash_2016_Going%20online%20on%20behalf%20of%20others_WEB.pdf.

Tolmie, Peter, Andy Crabtree, Tom Rodden, Chris Greenhalgh, and Steve Benford. 2007. "Making the Home Network at Home: Digital Housekeeping." *ECSCW*, London. https://doi.org/10.1007/978-1-84800-031-5_18.

van Dijck, José. 2020. *The Digital Divide*. Cambridge: Polity Press.

Watson, Ash, Deborah Lupton, and Mike Michael. 2020. "Enacting Intimacy and Sociality at a Distance in the COVID-19 Crisis: The Socio-materialities of Home-based Communication Technologies." *Media International Australia* 178, no. 1: 136–50. https://doi.org/10.1177/1329878X20961568.

Watson, Ash, Deborah Lupton, and Mike Michael. 2021. "The COVID Digital Home Assemblage: Transforming the Home into a Work Space During the Crisis." *Convergence* 27, no. 5: 1207–21. https://doi.org/10.1177/13548565211030848.

Wilding, Raelene. 2006. "'Virtual' Intimacies? Families Communicating Across Transnational Contexts." *Global Networks* 6, no. 2: 125–42. https://doi.org/10.1111/j.1471-0374.2006.00137.x.

Wilding, Raelene, Shashini Gamage, Shane Worrell, and Loretta Baldassar. 2022. "Practices of 'Digital Homing' and Gendered Reproduction among Older Sinhalese and Karen Migrants in Australia." *Journal of Immigrant & Refugee Studies* 20, no. 2: 220–32. https://doi.org/10.1080/15562948.2022.2046895.

Wilson, Tom, Peter McDonald, Jerome Temple, Bianca Brijnath, and Ariane Utomo. 2020. "Past and Projected Growth of Australia's Older Migrant Populations." *Genus* 76, no. 20: 1–21.

Worrell, Shane. 2021. "From Language Brokering to Digital Brokering: Refugee Settlement in a Smartphone Age." *Social Media + Society* 7, no. 2: 1–11. https://doi.org/10.1177/20563051211012365.

Yu, Haiqing, Ge Zhang, and Larissa Hjorth. 2023. "Mobilizing Care? WeChat for Older Adults' Digital Kinship and Informal Care in Wuhan Households." *Mobile Media & Communication* 11, no. 2: 294–311. https://doi.org/10.1177/20501579221150716.

"Zoom Saved All Our Lives"

A Case of Nonprofit Resilient Organizing during the COVID-19 Pandemic

Evgeniya Pyatovskaya, University of South Florida

Introduction[1]

The Covid-19 pandemic created a global crisis (Frosh and Georgiou 2022), and like any crisis of such scale, it impacted the most vulnerable populations in the United States (Kantamneni 2020). It also caused a "pedagogical disruption" (Yellow Horse and Nakagawa 2020, 353); opportunities for new ways to approach instruction and content emerged, while at the same time barriers arising from the unreflective reproduction of institutional arrangements and marginalizing practices flourished (Levinson and Markovits 2022). While educational institutions of all levels were swiftly closing their doors and shifting instruction online, nonprofit organizations (NPOs) that engaged in educational activities were brainstorming how to continue supporting their target populations in a quickly changing environment (Sacks and Jones 2020).

NPOs, like metaphorical platelets, rushed to help patch fresh societal wounds while transforming themselves and their practices to ensure that they could withstand the blows of the pandemic (Newhouse 2022). According to the State of the Nonprofit Sector Survey (Nonprofit Finance Fund 2022) 88 percent of NPOs admitted the way they work changed. Using Zoom to work and conduct programming remotely became one of such changes in NPO work routines (Newby and Branyon 2021).

This chapter will examine how New York City (NYC)-based NPO Summit (pseudonym) utilized Zoom as the platform of choice to continue its programming, support its youth, and expand its services. The purpose of this case study is to describe how the societal changes brought on by the pandemic within the first six months of its development (March 2020–September 2020) invoked resilience processes at Summit and, simultaneously, exposed inequities experienced by Asian immigrant (AI) and Asian American (AA) youth that had existed in educational processes long before the pandemic. It sheds light on issues of access and privilege that originate at the intersection of race, immigrant status, age, and language and were made more visible when online interaction via Zoom became a mandate. In effect, the use of Zoom (re)created a complex matrix of oppression embedded in the culture and society of the United States.

The focus on this organization is not accidental. New York was one of the states most severely affected by the pandemic and its deleterious side effect: increased discrimination and marginalization of specific groups. Counter to New York state then governor Andrew Cuomo's (2020) Tweet, where he called the pandemic "the great equalizer," meaning that the virus did not spare anyone regardless of their race, social class, and gender, Covid-19 had disproportionate effects on groups that have been systematically disadvantaged and oppressed in the United States. Yet, Summit was successful in shifting to Zoom-enabled programming and continued its services while 85 percent of NPOs reported program/event cancellations because of the pandemic.

The hierarchy of the US society that divides people based on class, race, gender, ethnicity, immigration status, and other identities, which Wilkerson (2023) refers to as a caste system, became clearly observable with the onset of the pandemic. Individuals' experiences of the pandemic varied based on their employment status, type of job, gender, race, ethnicity, and sexuality (Hansman 2022). Changes in daily routines led to some vulnerable groups experiencing heightened levels of discrimination and inequity. For example, in comparison with other marginalized groups, AIs and AAs suffered immense negative impacts from racist microaggressions, verbal attacks, and physical violence from being scapegoated in the media for the onset of the pandemic (see Choi and Lee 2021; Gover et al. 2020; Vachuska 2020). As pandemic stressors and questions about identity weighed on AI and AA individuals in NYC, the need for virtual spaces that could offer social connections, support for exploring and expressing identities, and mental health assistance became increasingly pressing (Liu 2022). NPOs that offered such programming became vital for this community's resilience.

Through reflexive thematic analysis (Braun and Clarke 2006, 2019, 2021, 2022) of two interviews with a manager responsible for implementing Zoom for the organization's elementary Saturday program, and autoethnographic personal reflections of the author who worked at the organization during the first six months of the pandemic as the second manager of the Saturday program, this chapter will focus on how multiple marginal positionalities of the AI and AA families the organization served intertwined and became visible in the complex process of resilience in the time of the pandemic. The quotes used in the chapter are all taken from the two interviews with the second manager of the program—Aileen (pseudonym).

The discussion surrounding racism, marginalization, and equity frequently overlooks AIs and AAs. The model minority myth (MMM) plays a central role in this omission. MMM imposes a set of cultural expectations on AA and AI individuals—for example, that they are naturally smart, wealthy, spiritually enlightened—and often blinds people to systemic struggles that AA and AI face. The MMM is insidious as it does not simply create a certain homogenous image of a very diverse population but also perpetuates unjust racial hierarchies, masks oppression that AA and AI experience, and limits state support to these groups (Yi et al. 2023). Instead, they are often perceived as having unproblematically achieved great success socially, economically, and educationally (Chen et al. 2021). This chapter serves as a starting point for initiating discussions on how Zoom and Zoom-based educational activities can be implemented mindfully to equitably address the needs of marginalized youth, including AI and AA youth, who are particularly susceptible to the consequences of the Covid-19 pandemic.

Summit

Summit (pseudonym) is a New York-based NPO that provides opportunities for low-income and underserved AA, AI, and other immigrant youth. It does so through offering mentoring, athletics, and elementary programs as well as college preparation and mental health programs through partnerships with NYC schools. The organization celebrated its thirty-first anniversary in 2023 and currently serves over 1,600 students and employs 42 individuals.

This chapter focuses on Summit's elementary Saturday program run at one of the largest elementary schools located in an immigrant neighborhood in Brooklyn. The program, established at the beginning of 2019 and run by Summit on the school's premises, quickly grew from 30 to 230 students in grades

2 through 5. By the time the pandemic began, the program was in the middle of its spring semester. It engaged seventy-seven volunteers (predominantly AI and AA, with a few being Latinx or white) and two full-time staff who managed the program on-site (Aileen and myself). Saturday sessions consisted of an academic component, where students worked with volunteers to improve their math and English skills, and an athletics program with basketball and yoga sessions. During the academic component, Summit sought to help the youth perfect their academic skills and provide positive role models to children through their interactions with volunteers. Each volunteer was assigned to work with two or three children consistently with whom they gradually built relationships.

The neighborhood's racial and ethnic composition is complex. About 35 percent of the residents are Asian and Latinx individuals. Therefore, most of the students at the school were AI or AA from various ethnic communities (e.g., Chinese, Korean, Filipino) and approximately 31.5 percent were Latinx. Children of Arab or Middle Eastern origin comprise about 4 percent of the school's population. In its mission, Summit commits to serving Asian and immigrant youth, which thereby allows it to include children of ethnicities other than Asian into its programs.

Summit's programming was severely affected by the school closure, mainly because of the looming uncertainty as to how the organization would continue to connect with the youth it served. The urgency of resolving the situation of the severed access to children was additionally warranted by the organization's serving one of the most vulnerable populations at the time of Covid-19: AI and AA youth. These youth not only suffered from all the adversities of the pandemic but also experienced heightened levels of anxiety and fear connected with growing anti-Asian hate (Akiba 2020; Exner-Cortens 2022). School closures and the transition to online learning exacerbated the stress as it made it difficult for the youth to connect with peers and teachers in person, seek advice, and be in community with those who went through similar experiences. While race, social class, and gender have been proven to result in disparities in access to education and educational results (Flores et al. 2019), the pandemic seriously intensified these disparities.

The disruption of the educational process caused by the school closure led Summit to seek alternatives to the modality of the Saturday program so that it could fulfill the deficiency in educational services and mental and emotional support needed to assist Asian and immigrant youth through the pandemic. A decision was made to immediately shift the program to Zoom.

It took Summit's two managers about two weeks in coordination with the school's administrators to identify children who needed the most help academically and emotionally (new immigrants, those who were falling behind in class before the pandemic, etc.), acquire families' contact information, reach out to those families, find volunteers who agreed to continue volunteering online, and design a new program to be run on Zoom.

Personal Reflections

When the pandemic struck, I worked as one of the two managers of the Summit's Saturday program. Aileen and I pioneered the program at the beginning of 2019 and scaled it up in the fall of 2019. We created the curriculum and made sure that the program ran smoothly every Saturday by coordinating the volunteer recruitment as well as student attendance, intake, and dismissal of students.

By the beginning of the spring semester in 2020, I knew many of the students by name and could recognize their parents when they came to pick them up. When the school closed, it left me anxious: the task of shifting our whole program online, of almost 300 people inclusive of the youth and volunteers, seemed insurmountable.

Communication Theory of Resilience

Resilience is a popular but not cohesive concept. It is generally defined as an ability to overcome a disruption. Alternative conceptualizations view resilience as a process and/or an outcome of individual or collective efforts to navigate disruptions (Buzzanell et al. 2009; Houston 2014).

In this chapter, I adopt the communicative view of resilience formulated by Patrice Buzzanell (2010, 2018) in the Communication Theory of Resilience (CTR). According to CTR, resilience in organizations "encompasses the process by which individuals and organizations reintegrate and foster productive change during and after career setbacks, material and personnel losses, disasters, or other obstacles" (Buzzanell 2018, 14). It is realized through five communicative processes: (1) crafting normalcy, (2) affirming the most meaningful identity anchors, (3) using and maintaining social networks, (4) finding alternative logics, and (5) acknowledging negative feelings and foregrounding positive action (Buzzanell et al. 2009).

In *crafting normalcy,* individuals engage in communication that serves to establish a set of patterns or customary ways of life, either directly or indirectly, that allow meaning to emerge (Buzzanell 2010). It includes adopting new behaviors, establishing new routines, and creating relationships. The second process is *affirming important identity anchors.* For example, in situations of job loss, individuals can anchor their important identities as parents or active community members. At the level of organizations, a crisis can inspire an organization to reaffirm its mission and values. *Maintaining and utilizing social networks* includes establishing, maintaining, and relying on relationships to cope with challenging circumstances. Cultivating resilience through a*lternative logics* means acknowledging, reframing, and/or making sense of the current circumstances even if sometimes in "ironic, contradictory, paradoxical, and even nonsensical ways" (Buzzanell 2010, 6). For example, viewing a disruption as an opportunity would be an example of putting alternative logics to work. Finally, the communicative process of *foregrounding positive action and backgrounding negative feelings* entails recognizing adversity and acknowledging one's right to experience negative emotions, while simultaneously concentrating on positive aspects and preserving a sense of hope.

The resilience of NPOs has been studied sparsely. Yet, NPOs—whose work is guided by existing social inequities and which are, therefore, entangled in unjust systems producing marginalization and vulnerability—must be resilient for organizational survival and continued support of the populations they serve. Resilience for NPOs becomes a mandate and a duty, rather than a choice. Building upon the framework set out by CTR, I contemplate how resilient processes that lead to generally positive outcomes for both the nonprofit and its target population make visible the very inequalities that triggered resilience.

Intersectionality and Resilience

The concept of intersectionality refers to the interdependence of race, gender, and class, which creates a complex matrix of discrimination and social inequities that exacerbate several societal problems, including health disparities and inequities in employment and educational opportunities (Crenshaw 2014). Certain historically oppressed groups experience the most extreme burdens of societal disruptions like the pandemic (Bowleg 2020; Maestripieri 2021). CTR posits that resilience does not solely depend on an individual or a group of individuals but also on material and social resources available to them (Buzzanell 2018; Buzzanell and Houston 2018). Thus, it is essential to examine how resilience is impacted by and impacts the state of oppression and marginalization.

Research that examined resilience through CTR has focused on various populations (e.g., see: Kuang et al. 2021; Long et al. 2015; Scharp et al. 2020, 2021; Tian and Bush 2020) including immigrant youth (e.g., Kam et al. 2018) to show how the enactment of resilience is dependent on socioeconomic and structural positionalities of individuals. A significant sign of structural inequity is education inequity. Discrimination and oppression that originate at the intersections of race, class, gender, social, and immigrant status create problems with academic adjustment and progress for AI and AA youth (e.g., Kiang et al. 2016). Stressors introduced by the pandemic were different "across students' social locations at the intersection of race/ethnicity, class, gender, immigration/ international student status" (Yellow Horse and Nakagawa 2020, 356) and amplified the negative impact of intersectional oppression in education. It is important, therefore, to view the resilience prompted by pandemic-connected stressors and historic oppressions, within an intersectional framework to move away from its idealization as a positive and unproblematic process. Intersectionality reveals resilience as gendered, classed, and racialized, with individuals of lower socioeconomic standing and people of color needing to engage in resilience processes more often than their white counterparts.

In the following sections, I discuss interactions and programming that aligned with all five communicative processes of CTR (Buzzanell 2010, 2018) and how these processes spotlighted the issues and inequities that had existed for a long time before the Covid-19 lockdowns. Specifically, Zoom is discussed as a modality that allowed Summit and its participants to craft normalcy and develop alternative logics of service provision and productive action that happened alongside organization-wide fears about the pandemic. The use of Zoom also uncovered the depth of the existing digital divide, the language barrier that persists online, and other issues of access and privilege.

Zoom Is the New Normal

The general vibe during this time was a lot of uncertainty, a lot of putting out fires, a lot of troubleshooting, and just kind of building the plane as we flew.

(Aileen, pers. comm.)

Online education was not invented with the onset of the pandemic; it has long been a supplement to in-person instruction or used on its own for its flexibility and the opportunities for inclusion of students regardless of their geographic location (Masry-Herzallah and Stavissky 2021). However, the shift to online modalities in education and education-related programming was abrupt and fast with the beginning of the pandemic (Moser et al. 2021). Schoolchildren were suddenly locked in their homes and told to attend lessons online (City of New York 2020). Such a momentous shift was neither planned nor carefully prepared for, neither by educational institutions and NPOs that worked with the youth nor by families and the youth themselves.

Summit paused its programming for two weeks to regroup and shift online. For the Saturday program, it meant first coordinating with the school to choose students to be invited into the online program and gathering families' contact information. Then, the two managers reached out to the parents by phone to invite their children to participate and explain the logistics of online programming.

For Summit, turning to Zoom was a way to create normalcy amidst chaos. It helped ensure uninterrupted programming and the ability to engage with the youth, which was an achievement at a time when face-to-face interactions were restricted. A speedy transition to Zoom also conferred a sense of normalcy to the youth as they were able to still "have Summit," be in community with their peers and volunteers, and continue learning.

The resilience process of creating normalcy had complexities, however. For those inside the organization, learning how to use Zoom and coming to terms with it as the new way to communicate and work was not an easy accomplishment. The changes that utilizing Zoom brought on also required a certain level of preparedness from the students so that they could take advantage of this new normal: (1) availability of devices and access to the internet and (2) digital literacy.

Availability of Devices

The digital divide presents a significant concern regarding equity, marking both a digital literacy gap and a device gap. Being digitally literate is unimportant if an individual does not have access to a digital device. As soon as the Saturday online program was launched, Summit faced the problem of the lack of devices among the students who signed up to participate. The digital device gap was ubiquitous for NYC families—nearly 300,000 students needed a device to participate in online learning (Kirby 2021). The Department of Education (DoE) responded to this disparity by distributing a similar number of devices with data plans to students (Hicks and Raskin 2021).

In line with a general trend, many Summit families reported experiencing delays in receiving the devices or not receiving a device at all (Kirby 2021). In September 2020, the Summit's partner school stated that 77 percent of their students had no access to the internet or had not received the devices they ordered (Cruz 2020). Given that most students attending the school were Asian and Latinx, it proved that the pandemic resulted in unequal access to educational resources, with varying impacts on different racial groups (Francis and Weller 2020). Immigrant communities and communities of color were disproportionately affected by the absence of both internet access and devices. Not surprisingly, delays in accessing the online program led to decreased motivation to participate in it and, therefore, fewer opportunities for youth to gain digital literacy.

As the online program developed, some children borrowed cell phones from their siblings and parents to participate in the program. Apart from the lack of devices, sharing can be explained by the fact that Asian families are usually larger than an average US family with several generations living together, which forces students to share devices and leads to limited online engagement (Dinh et al. 2020). Given that a student typically requires "a desktop computer, laptop, or tablet in their household" (Chandra et al. 2020, 8) to meet the minimum device threshold for online learning, AI and AA students of Summit suffered a severe device gap. Some of them simply did not show up for the online Saturday program.

Digital Literacy

According to Beaunoyer et al. (2020), there are disparities among individuals and social groups concerning both access to technology and their ability to derive benefits from its use. The children's and parents' digital literacy or lack thereof became one of the first and most serious obstacles Summit faced when transferring the program online. Program managers spent hours in phone conversations with the parents explaining how to install Zoom, login, and join a Saturday session.

Research has shown that foreign-born parents' digital literacy is low (Cherewka 2020). In line with these findings, most of the parents with whom

Summit managers spoke had issues understanding how Zoom worked. One of the biggest potential risks that arose from the parents' lack of digital literacy was an inability to facilitate their children's participation online, younger children's failure to learn the technology by themselves, and children's consequent failure to participate in online programming. Another issue was the parents' lack of digital skills to troubleshoot *during* the online session in situations of lost connection, sound, or video, which also precluded students' successful participation in the program. As a result, several children were excluded from the online program.

Introducing Zoom as an alternative to in-person programming inevitably revealed digital exclusion, which is a type of social marginalization that exacerbates both material and social deprivation (Beaunoyer et al. 2020). Importantly, being digitally excluded leads to precarity in such crucial areas as education and social connections. In the case of Summit, the introduction of Zoom to establish a new normal and promote resilience led to the realization of digital inequality endemic to AI and AA families. As immigrants generally have lower levels of digital proficiency compared to those adults who speak the native language (Bergson-Shilcock 2020), it became an obstacle to the successful delivery of the program to all children who were admitted. Even when the students gained access to remote learning, it was not always effective.

Affirming Identity Anchors

Individual values and mission are very much
rooted in what we're trying to do together as a group.

(Aileen, pers. comm.)

The pandemic brought very specific and powerful stressors to the lives of AIs and AAs. Anti-Asian hate in New York City grew exponentially as the pandemic ran rampant, partially owing to malicious remarks made by President Trump, who referred to Covid-19 as the "Chinese virus" (Fallows 2020), and Mike Pompeo, Secretary of State, who chose "Wuhan virus" (Levenson 2020) instead. Physical abuse and violence against AIs and AAs in public places like grocery stores, public transportation, and simply on the streets (Jeung and Nham 2020) soon followed. Anti-Asian hate put the very identities of students and Summit employees under attack.

Buzzanell (2010) points out that "collectivities, and the individuals embedded within, construct and reinforce particular, pivotal identities that operate as identity anchors" (5). The anxiety of being and/or looking Asian during the Covid-19 pandemic prompted resilience through affirming an identity anchored in being an AI and/or AA within the organization and in communication with the students. The more news appeared about anti-Asian crimes, the more active Summit was in brainstorming new sessions and other ways they could be helpful to the youth. As a result of continuously engaging with the common identification frame of the unity of AI and AA community and Summit's role in it, well-being and mental health sections that focused on questions of identity and belonging were introduced into the Saturday program.

Summit used Saturday gatherings on Zoom in part to inquire about students' well-being. It also allowed a safe gathering space for the students, staff, and volunteers where all could be in a community with people who were similar in some way, experiencing similar anxieties and fears, and understanding each other's struggles. The Zoom classroom became a healing place where intersecting discourses of togetherness and similarities of struggles worked to elevate and enhance an important identity.

At the same time, limitations connected to Zoom use soon became prominent in the realization of this resilience process, including: (1) the constant need to overcome a language barrier and (2) Summit's limited capacity to run its Zoom-based program.

The Language Barrier

According to Summit's website, one in every five AA youth in New York City lives in a household where no one older than fourteen speaks English well, or at all. Among the students chosen by the school administration to participate in the Saturday program, there were two students who had only recently immigrated to the United States and were English language learners. The language barrier did not allow these students to participate meaningfully in whole group activities. However, a solution was soon found, and a dedicated volunteer was assigned to each of them for personal attention to their language learning needs. The students were able to participate in the group gatherings accompanied by their volunteers.

The more serious issue that Zoom illuminated was the parents' lack of English proficiency. Kerker et al. (2021) confirm that during the pandemic "immigrant-origin communities faced additional challenges, including language barriers to accessing information and limited access to most forms of federal aid or federal public benefits" (271), which points to deep-seated inequities.

To address the language barrier problem, DoE ran an over-the-phone interpretation program where parents who did not speak English could get information about their child's education interpreted to them in real time. Summit utilized the service using the schools' credentials to talk to parents and invite their children to the programs, coordinate participation, and troubleshoot Zoom issues. Connecting with parents who did not speak English through interpreters opened a door to Summit's knowledge of other barriers that existed in these families' lives. Some parents, unable to seek answers, help, or advice anywhere else, confided in Summit managers. For example, a mother who suffered a miscarriage asked for help getting groceries. Other requests revolved around questions about virtual learning, devices, and extra tutoring. Limited language ability disproportionately and directly impacted students who participated in Zoom-based learning but also indirectly hindered their opportunities for participation because of their parents' lack of English language command and inaccessibility of interpretation services.

Limited Capacity

Zoom became a way to continue serving the youth Summit served prior to the pandemic; however, it soon became clear that it was impossible to accommodate the same number of students on Zoom. One major factor was the attrition of volunteers who did not want to engage, or could not engage, in virtual volunteering. The reasons differed, including: "Zoom fatigue"—lack of perceived "real" connection with the students and not liking Zoom as a medium for interaction. Fewer volunteers limited Summit's capacity to serve the youth. Therefore, only students who were identified as "high need" academically and/ or socially-emotionally by the school administration were admitted into the Saturday program.

Even though the Zoom online classroom allowed hosting of many participants, managing an online program with over 100 attendees was not possible given Summit's decreased capacity to rely on volunteers, which, consequently, impacted the organizational capacity to enroll more students.

Resilience cultivation through affirming common identities was limited in more than one way to a select group of students.

Maintaining and Using Social Networks

Maintaining and using social networks was the third process through which resilience was actualized. The networks were located on two levels: interorganizational, where Summit leveraged the resources inherent in social relationships between organizations (Buzzanell 2010); and a network between the organization and the families it served.

To alleviate the stressors of the pandemic, maintaining a social network that students had with Summit and its volunteers prior to the pandemic provided a sense of stability and community. It was especially important since AI and AA students experienced the highest decline in emotional health among other groups (Margolius et al. 2020), yet had trouble reaching out for support partially because of the MMM, which depicts AIs and AAs as a monolithic group that is hardworking and has triumphed over adversity, oppression, and discrimination (Chen et al. 2021). Such framing not only conceals the incidents of discrimination faced by young AIs and AAs but also makes them fear letting other people down (Chen et al. 2021; Gupta et al. 2011). Zoom became a safe space where mental health conversations in community with others were normalized and encouraged.

Margolius et al. (2020) argued that during the pandemic "students [experienced] a collective trauma and that they and their families would benefit from immediate and ongoing support for basic needs, physical and mental health, and learning opportunities" (1). For parents, especially those who lacked English proficiency, Summit became an organization that partially provided such support and an entry point to a larger social network (Summit's partner organizations and the school administrators) that they could leverage in situations where assistance was needed. Summit's interaction with the parents grew as Zoom was introduced as a platform for the Saturday program, and the organization became privy to the difficulties that families experienced outside of their children's learning journeys. In response, Summit was able to leverage its networks to alleviate some of the issues that families faced. For example, Summit's Saturday program managers obtained and distributed 200 coupons for a free meal to families of the children attending the program.

The coupons came from a partner organization whose employee was also a volunteer at Summit and knew firsthand about the difficulties in getting nutritious meals that AI and AA families experienced at the height of the pandemic.

Maintaining a connection with the parents played an important role in cultivating the resilience of the families it served. The organization's ongoing presence in these families' lives, and its readiness to assist in the areas where it saw need, and act as the families' advocate outside of the realm of education, positively impacted the families' journey through the pandemic. However, there was a degree of remoteness that Zoom was unable to overcome.

Insuperable Remoteness

As previously discussed, DoE provided families, including the ones that Summit served in its Saturday program, with devices and data plans. However, not all the families that needed them received the devices. An additional issue with equitable access to data surfaced when parents shared that even though they received a device, the location where the family lived did not have Wi-Fi or a broadband connection, rendering the device useless. Summit fundraised to obtain hotspots for families that did not have any other way to access the internet. Yet the problem was indicative of a larger societal issue where "much of the recovery effort in the U.S. left immigrant origin families behind, and as a result, these families struggled further to meet basic needs during the pandemic" (Kerker et al. 2021, 271) including making sure that children have access to educational opportunities.

Finding Alternative Logics

Exploring options and being creative was part of the process of cultivating resilience by adopting alternative logics of Zoom as a classroom with a twist. Zoom allowed the organization to keep working, rather than to temporarily suspend programming or close its doors completely. In a way, Zoom helped explore and test a novel way of working that the staff might have thought about prior to the pandemic, but without the crisis it would have taken much longer for Summit to test the online modality of its programming. Alternative logics of Zoom sped up programming and organizational transformation as program managers maneuvered between the usual and new work practices to ensure Zoom sessions were as productive as in-person ones.

Summit engaged in resilient thinking by first acknowledging that the old ways of doing programming were no longer available and that the organization had no control over the timeline of the pandemic. Through conversations during staff meetings, the staff imagined what a Zoom-based program could look like and kept working through various ideas. The program manager said: "It was a lot of trial and error. I feel like because this was the first time for all of us doing it. I learned the hard way from a lot of things of what not to do, and what to do going forward" (Aileen, pers. comm.) The creative approach, persistence, and flexibility of the staff and management made the transition to Zoom possible and allowed staff to retain a commitment to the organization's general goal of helping immigrant and Asian youth.

Zoom embodied resilience through alternative logics. Yet, Zoom-based program highlighted inequity in its other form—unattended children.

Children Unattended

Attending Zoom sessions looked very different for students of different socioeconomic status. As one of the participants shared:

> This whole Zoom thing and being online brought up a lot of things that you normally wouldn't think about. For example, we were able to see . . . the kids' living situations. Things that I noticed were students' rooms . . . and you could see that maybe three or four of them are crammed into one little bedroom, with bunk beds and things like that. . . . Or even the noise level, right? How chaotic it is. How does a student focus in that kind of environment when there's like a crying kid in the background and people screaming?

The Zoom screen revealed a lot of instances where inequity in job opportunities and education was strikingly obvious. Remote learning added an additional burden on families since it necessitated significant adult participation in children's daily educational activities (Kerker et al. 2021). However, many AI and AA parents were engaged in jobs that could not be performed remotely (e.g., food preparation, construction, transportation) (Gemelas et al. 2021), thereby limiting their ability to assist their children with learning Zoom. Children whose parents could not work from home during the pandemic, especially in the beginning, skipped more sessions than others and could not participate as productively due to connectivity and other issues that they were unable to troubleshoot by themselves.

Another obstacle to student engagement with the Zoom-based programming was crowdedness of the students' homes, as noted by one of the managers in the earlier quote. Summit's experience with running a Zoom-based Saturday program confirmed Kerker et al.'s (2021) observation that families with multiple children faced greater difficulties in remote learning since setting up an environment where all the children could concentrate simultaneously in the same household was often a significant challenge.

Legitimizing Negative Feelings While Foregrounding Productive Action

Finally, during such an uncertain and highly stressful time as the first six months of the pandemic, Summit organizational members talked about intentionally focusing on productive actions and next steps while allowing space to vent about the fears and anxieties that related not only to their work but also their personal lives. Summit's employees had one-on-one meetings with their supervisors where they could discuss feelings stirred up by the pandemic and acknowledge their fears, anger, and frustration.

Zoom was a place for happy hours that allowed peers to talk with each other about things and happenings unrelated to the work they were doing. They also allowed staff to maintain a positive outlook on the situation and give and receive support from peers. Zoom became a space of support not only for the students but also for the staff and a place where negative feelings were legitimized, and verbal support and encouragement were offered.

In terms of the Saturday program, Zoom played a similar role—a place where students were able to talk about their own frustrations or share a laugh with a volunteer. Art projects and group work that happened over Zoom gave students a sense of productive work and a positive boost from interacting with peers and volunteers about important topics, letting them forget about the pandemic for at least the duration of a session.

Conclusion

To conclude, organizational resilience is cultivated through several communicative processes that, taken together, enhance an organization's ability to transform during a crisis, weather it, and emerge on the other side stronger and renewed. Resilience processes are neither linear nor dependent on each other: context, the severity of the disruption, and organizational preparedness all make specific communicative resilience processes manifest stronger at certain points in time. However, each one enables an organization's survival.

Communicative resilience is a collective process, and it does not always serve the interests of everyone involved in it (Buzzanell 2010, 2018). In the case of NPOs, where organizational attempts at resilience have high stakes for populations they serve, it is important to be attentive to how some groups and/ or individuals may be excluded from deriving the benefits of certain resilient moves. During the pandemic, AI and AA populations served by Summit had limited access to technology due to a number of confounding factors like poverty, living in multigenerational households, and limited English proficiency, which precluded some of the students from participating in Saturday online programming, making apparent that access to Zoom was far from equal.

Zoom was not a new videoconferencing tool; it was the specificity of the situation that made it somewhat of an innovative solution and the one that enabled resilience and served as a springboard and a tool for its actualization. As a NPO with a commitment to serving immigrant and Asian youth in New York City, Summit actively worked to not only maintain organizational integrity and operations but also address and resolve the issues stemming from unequal access to technology that originated from overlapping marginalized positions of Summit's target populations. Zoom served as a litmus test in identifying the most pressing of these issues, yet intensified marginalization of those participants who did not have equal access to technology or did not have the knowledge and/or skills to operate it.

It is our collective responsibility to stay vigilant to the inequities that are often invisible yet destructive to individual and collective prosperity and success. Communication can improve the well-being and resilience of individuals, communities, and organizations during disruptions, and it is important that in our communicative resilience journey, we proceed with care, attention, and preparedness to act radically to address the inequities that become palpable because of this process.

Note

1   This work was supported by the Waterhouse Family Institute for the Study of Communication and Society's grant [20230102].

References

Akiba, Daisuke. 2020. "Reopening America's Schools during the COVID-19 Pandemic: Protecting Asian Students from Stigma and Discrimination." *Frontiers in Sociology* 5: 1–6. https://doi.org/10.3389/fsoc.2020.588936.

Beaunoyer, Elisabeth, Sophie Dupéré, and Matthieu J. Guitton. 2020. "Covid-19 and Digital Inequalities: Reciprocal Impacts and Mitigation Strategies." *Computers in Human Behavior* 111. https://doi.org/10.1016/j.chb.2020.106424.

Bergson-Shilcock, Amanda. 2020. "The New Landscape of Digital Literacy." *National Skills Council*, May 20, 2020. https://nationalskillscoalition.org/wp-content/uploads /2020/12/05-20-2020-NSC-New-Landscape-of-Digital-Literacy.pdf.

Bowleg, Lisa. 2020. "We're Not All in This Together: On Covid-19, Intersectionality, and Structural Inequality." *American Journal of Public Health* 110, no. 7: 917–17. https:// doi.org/10.2105/ajph.2020.305766.

Braun, Virginia, and Victoria Clarke. 2006. "Using Thematic Analysis in Psychology." *Qualitative Research in Psychology* 3, no. 2: 77–101. https://doi.org/10.1191 /1478088706qp063oa.

Braun, Virginia and Victoria Clarke. 2019. "Reflecting on Reflexive Thematic Analysis." *Qualitative Research in Sport, Exercise and Health* 11, no. 4: 589–97. https://doi.org /10.1080/2159676x.2019.1628806.

Braun, Virginia and Victoria Clarke. 2021. "To Saturate or Not to Saturate? Questioning Data Saturation as a Useful Concept for Thematic Analysis and Sample-Size Rationales." *Qualitative Research in Sport, Exercise and Health* 13, no. 2: 201–16. https://doi.org/10.1080/2159676x.2019.1704846.

Braun, Virginia and Victoria Clarke. 2022. "Toward Good Practice in Thematic Analysis: Avoiding Common Problems and Be(com)ing a *Knowing* Researcher." *International Journal of Transgender Health* 24, no. 1: 1–6. https://doi.org/10.1080 /26895269.2022.2129597.

Buzzanell, Patrice M. 2010. "Resilience: Talking, Resisting, and Imagining New Normalcies into Being." *Journal of Communication* 60, no. 1: 1–14. https://doi.org/10 .1111/j.1460–2466.2009.01469.x.

Buzzanell, Patrice M. 2018. "Organizing Resilience as Adaptive-Transformational Tensions." *Journal of Applied Communication Research* 46, no. 1: 14–18. https://doi .org/10.1080/00909882.2018.1426711.

Buzzanell, Patrice M., and Brian J. Houston. 2018. "Communication and Resilience: Multilevel Applications and Insights – A *Journal of Applied Communication Research* Forum." *Journal of Applied Communication Research* 46, no. 1: 1–4. https://doi.org /10.1080/00909882.2017.1412086.

Chandra, Sumit, Amy Chang, Lauren Day, Amina Fazlullah, Jack Liu, Lane McBride, Thisal Mudalige, and Danny Weiss. 2020. "Closing the K-12 Digital Divide in the Age of Distance Learning." *Common Sense Media*. https://www.commonsensemedia

.org/sites/default/files/featured-content/files/common_sense_media_report_final_7_1_3pm_web.pdf.

Buzzanell, Patrice M., Suchitra Shenoy, Robyn V. Remke, and Kristen Lucas. 2009. "Responses to Destructive Organizational Contexts." In *Destructive Organizational Communication: Processes, Consequences, and Constructive Ways of Organizing*, edited by Pamela Lutgen-Sandvik and Beverly Davenport Sypher, 293–315. New York: Routledge.

Chen, Szu-Yu, Tzu-Fen Chang, and Kristy Y. Shih. 2021. "Model Minority Stereotype: Addressing Impacts of Racism and Inequities on Asian American Adolescents Development." *Journal of Child and Adolescent Counseling* 7, no. 2: 118–31. https://doi.org/10.1080/23727810.2021.1955544.

Cherewka, Alexis. 2020. "The Digital Divide Hits U.S. Immigrant Households Disproportionately during the COVID-19 Pandemic." *Migration Policy Institute*, September 3, 2020. https://www.migrationpolicy.org/article/digital-divide-hits-us-immigrant-households-during-covid-19.

Choi, Hee An and Othelia EunKyoung Lee. 2021. "To Mask or to Unmask, That Is the Question: Facemasks and Anti-Asian Violence during COVID-19." *Journal of Human Rights and Social Work* 6, no. 3: 237–45. https://doi.org/10.1007/s41134-021-00172-2.

Crenshaw Kimberlé. 2014. *On Intersectionality Essential Writings*. New York: New Press.

Cruz, David. 2020. "Some Kindergarten Students in Sunset Park Left without Devices for Remote Learning." *Gothamist*, September 19, 2020. https://gothamist.com/news/some-kindergarten-students-sunset-park-left-no-device-remote-learn-next-week.

Cuomo, Andrew (@NYGovCuomo). "This Virus Is the Great Equalizer." *Twitter*, March 31, 2020. https://twitter.com/nygovcuomo/status/1245021319646904320.

Dinh, Quyen T, Katrina D Mariategue, and Anna H Byon. 2020. "Covid-19 - Revealing Unaddressed Systemic Barriers in the 45th Anniversary of the Southeast Asian American Experience." *Journal of Southeast Asian American Education and Advancement* 15, no. 2: 1–9. https://doi.org/10.7771/2153–8999.1209.

Exner-Cortens, Deinera, Kelly D. Schwartz, Carly McMorris, and Erica Makarenko. 2022. "Stress among Asian Youth during COVID-19: Moderation by Educational, Spiritual, and Cultural Sources of Belonging." *Journal of Adolescent Health* 70, no. 3: 500–3. https://doi.org/10.1016/j.jadohealth.2021.10.007.

Fallows, James. 2020. "2020 Time Capsule #5: The 'Chinese Virus.'" *The Atlantic*, March 18, 2020. https://www.theatlantic.com/notes/2020/03/2020-time-capsule-5-the-chinese-virus/608260/.

Flores, Lisa Y., Leticia D. Martinez, Gloria G. McGillen, and Johanna Milord. 2019. "Something Old and Something New: Future Directions in Vocational Research with People of Color in the United States." *Journal of Career Assessment* 27, no. 2: 187–208. https://doi.org/10.1177/1069072718822461.

Francis, Dania, and Christian E. Weller. 2020. "The Black-White Wealth Gap Will Widen Educational Disparities during the Coronavirus Pandemic." *Center for American Progress*, August 12, 2020. https://www.americanprogress.org/ article/black-white-wealth-gap-will-widen-educational-disparities-coronavirus -pandemic/.

Frosh, Paul and Myria Georgiou. 2022. "Covid-19: The Cultural Constructions of a Global Crisis." *International Journal of Cultural Studies* 25, no. 3–4: 233–52. https:// doi.org/10.1177/13678779221095106.

Gemelas, Jordan, Jenna Davison, Case Keltner, and Samantha Ing. 2021. "Inequities in Employment by Race, Ethnicity, and Sector during COVID-19." *Journal of Racial and Ethnic Health Disparities* 9, no. 1: 350–5. https://doi.org/10.1007/s40615-021 -00963-3.

Gover, Angela R., Shannon B. Harper, and Lynn Langton. 2020. "Anti-Asian Hate Crime during the COVID-19 Pandemic: Exploring the Reproduction of Inequality." *American Journal of Criminal Justice* 45, no. 4: 647–67. https://doi.org/10.1007/ s12103-020-09545-1.

Gupta, Arpana, Dawn M. Szymanski, and Frederick T. Leong. 2011. "The 'Model Minority Myth': Internalized Racialism of Positive Stereotypes as Correlates of Psychological Distress, and Attitudes Toward Help-Seeking." *Asian American Journal of Psychology* 2, no. 2: 101–14. https://doi.org/10.1037/a0024183.

Hansman, Catherine A. 2022. "'We're All in This Together'? Reflections on Inequity during Tumultuous Times." *New Directions for Adult and Continuing Education* 2022, no. 173–174: 9–19. https://doi.org/10.1002/ace.20448.

Hicks, Nolan and Sam Raskin. 2021. "Doe Bungled Distribution of Student iPads during COVID: Comptroller Report." *New York Post*, July 28, 2021. https://nypost .com/2021/07/28/doe-bungled-distribution-of-student-ipads-during-pandemic/.

Houston, Brian J. 2014. "Bouncing Forward: Assessing Advances in Community Resilience Assessment, Intervention, and Theory to Guide Future Work." *American Behavioral Scientist* 59, no. 2: 175–80. https://doi.org/10.1177/0002764214550294.

Jeung, Russell and Kai Nham. 2020. "Incidents of Coronavirus-Related Discrimination - AAPI Equity Alliance." *Asian Pacific Policy and Planning Council*, April 23, 2020. https://www.asianpacificpolicyandplanningcouncil.org/wp-content/uploads/STOP _AAPI_HATE_MONTHLY_REPORT_4_23_20.pdf.

Kam, Jennifer A., Debora Pérez Torres, and Keli Steuber Fazio. 2018. "Identifying Individual- and Family-Level Coping Strategies as Sources of Resilience and Thriving for Undocumented Youth of Mexican Origin." *Journal of Applied Communication Research* 46, no. 5: 641–64. https://doi.org/10.1080/00909882.2018 .1528373.

Kantamneni, Neeta. 2020. "The Impact of the COVID-19 Pandemic on Marginalized Populations in the United States: A Research Agenda." *Journal of Vocational Behavior* 119: 1–4. https://doi.org/10.1016/j.jvb.2020.103439.

Kiang, Lisa, Melissa R. Witkow, and Taylor L. Thompson. 2016. "Model Minority Stereotyping, Perceived Discrimination, and Adjustment among Adolescents from Asian American Backgrounds." *Journal of Youth and Adolescence* 45, no. 7: 1366–79. https://doi.org/10.1007/s10964-015-0336-7.

Kirby, Wesley. 2021. "NYC Department of Education Response to the COVID-19 Pandemic." Office of the New York State Comptroller. September 2021. https://www.osc.ny.gov/files/reports/osdc/pdf/report-8-2022.pdf.

Kerker, Bonnie D., Natalia Rojas, Spring Dawson-McClure, and Cristina González. 2023. "Re-Imagining Early Childhood Education and School Readiness for Children and Families of Color in the Time of COVID-19 and Beyond." *American Journal of Health Promotion* 37, no. 2: 270–73. https://doi.org/10.1177/08901171221140641c.

Kuang, Kai, Zhenyu Tian, Steven R. Wilson, and Patrice M. Buzzanell. 2021. "Memorable Messages as Anticipatory Resilience: Examining Associations among Memorable Messages, Communication Resilience Processes, and Mental Health." *Health Communication* 38, no. 6: 1136–45. https://doi.org/10.1080/10410236.2021.1993585.

Levenson, Thomas. 2020. "Conservatives Try to Rebrand the Coronavirus." *The Atlantic*, March 11, 2020. https://www.theatlantic.com/ideas/archive/2020/03/stop-trying-make-wuhan-virus-happen/607786.

Levinson, Meira and Daniel Levinson Markovits. 2022. "The Biggest Disruption in the History of American Education." *The Atlantic*, June 23, 2022. https://www.theatlantic.com/ideas/archive/2022/06/covid-learning-loss-remote-school/661360/.

Liu, Marcia M. 2022. "Where We're Really From: NYC Asian American Students Navigating Identity, Racial Solidarity, and Wellness during a Pandemic." *AAPI Nexus: Policy, Practice and Community* 19, no. 1–2: 105–22.

Long, Ziyu, Patrice M. Buzzanell, Min Wu, Rahul Mitra, Kai Kuang, and Huijung Suo. 2015. "Global Communication for Organizing Sustainability and Resilience." *China Media Research* 11, no. 4: 67–77.

Maestripieri, Lara. 2021. "The COVID-19 Pandemics: Why Intersectionality Matters." *Frontiers in Sociology* 6: 1–6. https://doi.org/10.3389/fsoc.2021.642662.

Margolius, Max, Alicia Doyle Lynch, Elizabeth Pufall Jones, and Michelle Hynes. 2020. "The State of Young People during COVID-19: Findings from a Nationally Representative Survey of High School Youth." *America's Promise Alliance*, May 31, 2020. https://eric.ed.gov/?q=source%3A%22America%27s%2BPromise%2BAlliance%22&id=ED606305.

Masry-Herzallah, Asmahan, and Yuliya Stavissky. 2021. "Investigation of the Relationship between Transformational Leadership Style and Teachers' Successful Online Teaching during COVID-19." *International Journal of Instruction* 14, no. 4: 891–912. https://doi.org/10.29333/iji.2021.14451a.

Moser, Kelly M., Tianlan Wei, and Devon Brenner. 2021. "Remote Teaching during COVID-19: Implications from a National Survey of Language Educators." *System* 97: 1–15. https://doi.org/10.1016/j.system.2020.102431.

"New York City to Close All School Buildings and Transition to Remote Learning." 2020. *The Official Website of the City of New York*, March 15, 2020. https://www.nyc .gov/office-of-the-mayor/news/151-20/new-york-city-close-all-school-buildings -transition-remote-learning.

Newby, Kara and Brittany Branyon. 2021. "Pivoting Services: Resilience in the Face of Disruptions in Nonprofit Organizations Caused by COVID-19." *Journal of Public and Nonprofit Affairs* 7, no. 3: 443–60. https://doi.org/10.20899/jpna.7.3.443–460.

Newhouse, Chelsea. 2022. *Mission and More: The Economic Power of the Hudson Valley's Charitable Nonprofit Sector*. New York: New York Council of Nonprofits & Hudson Valley Funders Network.

Nonprofit Finance Fund. 2022."State of the Nonprofit Sector Survey." *Nonprofit Finance Fund*, 2022. https://nff.org/learn/survey.

Sacks, Vanessa and Rebecca M. Jones. 2020. "Nonprofit Organizations and Partnerships Can Support Students during the COVID-19 Crisis - Child Trends." *ChildTrends*, June 17, 2020. https://www.childtrends.org/blog/nonprofit-organizations-and -partnerships-can-support-students-during-the-covid-19-crisis.

Scharp, Kristina M., Devon E. Geary, Brooke H. Wolfe, Tiffany R. Wang, and Margaret A. Fesenmaier. 2020. "Understanding the Triggers and Communicative Processes That Constitute Resilience in the Context of Migration to the United States." *Communication Monographs* 88, no. 4: 395–417. https://doi.org/10.1080/03637751 .2020.1856395.

Scharp, Kristina M, Tiffany R Wang, and Brooke H Wolfe. 2021. "Communicative Resilience of First-Generation College Students during the COVID-19 Pandemic." *Human Communication Research* 48, no. 1: 1–30. https://doi.org/10.1093/hcr/hqab018.

Tian, Zhenyu and Hannah R. Bush. 2020. "Half the Sky: Interwoven Resilience Processes of Women Political Leaders in China." *Journal of Applied Communication Research* 48, no. 1: 70–90. https://doi.org/10.1080/00909882.2019.1704829.

Vachuska, Karl F. 2020. "Initial Effects of the Coronavirus Pandemic on Racial Prejudice in the United States: Evidence from Google Trends." *SocArXiv*. https://doi.org/10 .31235/osf.io/bgpk3.

Wilkerson, Isabel. 2023. *Caste: The Origins of Our Discontents*. New York: Random House.

Yellow Horse, Aggie J. and Kathryn Nakagawa. 2020. "Pedagogy of Care in Asian American Studies during the COVID-19 Pandemic." *Journal of Asian American Studies* 23, no. 3: 353–65. https://doi.org/10.1353/jaas.2020.0029.

Yi, Jacqueline, Raymond La, B. Andi Lee, and Anne Saw. 2023. "Internalization of the Model Minority Myth and Sociodemographic Factors Shaping Asians/Asian Americans' Experiences of Discrimination during COVID-19." *American Journal of Community Psychology* 71, no. 1–2: 123–35. https://doi.org/10.1002/ajcp.12635.

(Un)expected Errors

Mark Nunes, Appalachian State University, and
CassandraOzog, University of Regina

Throughout this collection of essays, we have highlighted how the logic of
Zoom operates within an ideological framework of adaptation, efficiency, and
maximized performance that likewise supports the network economies of
neoliberalism. But Zoom and other videoconferencing platforms are far from
monolithic in their uses and potentials. Several of the chapters in this volume have
also called attention to openings for alternate, "transverse" engagements within
these same apparatuses of control. The contributions of Howard, Cabalquinto,
and Pyatovskaya have brought to the fore the tactics of marginalized individuals
and communities whose position within these structures of control and
surveillance challenges the entrepreneurial subject "default settings" that define
and delimit the normative role of platform user. Other chapters in this collection
have offered accounts of how videoconferencing platforms such as Zoom afford
usages and modes of engagement that were not only unforeseen in design but
that also give rise to modes of practice that call into high relief the dominant
logic of institutional spaces (Grushka et al.; Cochrane & Kenney; Carmack et
al.), aesthetic practices (O'Brien; Fahner), and interpersonal interactions (Sci;
McArthur).

In this final chapter, we pick up on this line of thought by expanding upon a
point raised by Crano in the first chapter of this collection through a discussion
of how the logic of Zoom creates a "frame" that precedes the familiar framing
of subjects inside the Zoom window. This frame casts *stable connection* as
a performative norm and places an assumed universal user in a position of
panoptic regulation and control. In order to establish and maintain what
we might refer to as its ideological and technological "terms of service," this
frame must create its own "blind spot"[1] to the everyday, lived experience of
the platform's users. We challenge this frame—or at very least "call the frame

into question" in the hope of "show[ing] that the frame never quite contained the scene it was meant to limn, that something was already outside, which made the very sense of the inside possible, recognizable" (Butler 2016, 9). This framing that precedes the frame operates as much on its users as it does on the streaming content it attempts to contain, bracketing off the glitches, frozen images, and slow connection speeds—as well as the disruptive, unintended, and at times subversive uses of the platform—as excluded errors that are *handled*[2] and accounted for, but that ultimately do not signify. Such moments are kept out of frame as anomalous events within the technological promise of efficiency, performance, and a transcendence over space and time that not even a global pandemic could deter. Yet such moments are hardly outside the norm of our experiences on Zoom. In contrast to the ideological frame that excludes these aberrant encounters as somehow beyond the norms of operation, we argue that "unexpected" glitches and errors in performance and use reveal that Zoom's promise of a seamless, frictionless platform for work and sociality is built upon a network of precarious and unstable connections enacted between technological and social actors.

You're Muted

Zoom's performance logic casts out of frame errors that are very much a part of the everyday, lived experience of platform users. To start with the most common occurrence of a videoconferencing session gone awry, we present what became the unofficial slogan for life on Zoom during the pandemic:

"You're muted."

We have all been guilty of this videoconferencing misstep, and more likely than not, we have been equally guilty of calling out individuals who have likewise failed to perform "properly" within the social and technological protocols of the platform. The errant operator appears in their designated frame within the Zoom grid, lips moving, yet failing to communicate. Others on the call (often several) remind the (non)speaker that they have forgotten to unmute in an offer of help that is likewise a form of public shaming. The expression on their face says it all, as they come to realize that they have been talking into a void: the embarrassment, the frustration—the frantic, apologetic, and exaggerated gestures to try to communicate that they have heard the other participants and are trying to resolve the issue. Just as remaining muted introduces an unexpected error in the form of a stoppage of communication, so too does *failing to mute* present itself as an aberration in an otherwise friction-free exchange. Intrusive, unwanted noise does not merely interfere; it wrests control of the meeting itself, forcing background noise (in Speaker View at least) into the center frame. This precarious performance between signal and noise also occurs in the increasingly common hybrid videoconferencing spaces in which in-person participants and Zoom attendees interact with one another, forcing individuals to negotiate a technical dance between muting and unmuting required by their simultaneous role as both audience and presenter, caught in an endless feedback loop of reverb that reveals what has always been true: users are constantly receiving the same signal they send.[3]

These moments reveal both the illusion of seamless control built into the platform and the human performance that fails to operate by these protocols: if only one knew how to operate the controls properly, none of these errors would occur. But of course they do, and all the time. This "simple" act of knowing when and how to "mute" and "unmute" operates as a Taylorist measure of efficiency, and one that creates an overlap between professional behavior and interface design in which the seamless and frictionless encounter promised by the platform is contingent upon the digital fluency and technological grace of its participants. In contrast, we once again turn to Ahmed's (2006) reading of mastery, fluency, and success as a measure not of skill or prowess but of one's privileged encounter with a world that seems oriented toward one's own use (139).

These disorienting gaps in fluent use also demonstrate layers of privilege rooted in regular access to and use of technology. As the two preceding chapters in this section explore, these gaps in technological knowledge function as indices of precarity within a number of purposely marginalized groups including, but not limited to, immigrants and senior citizens. "Helpful" public shaming of a failure to perform efficiently within a platform's control logic skims over the often complex reasons why Zoom users struggle with videoconferencing tools, let alone the social expectations that accompany them. Should the user manage their privilege efficiently enough to master their controls, they may still encounter a technological precarity announced on screen through the dreaded words: "Your Internet Connection is Unstable." At such moments, a strange rupture occurs between the experience of the unstable user and everyone else on the call. For those on the call, the user freezes, and what ensues is another ritual of failure: the chant of "You're frozen!" Of course, this signal never arrives to the recipient, who may be well aware that they are frozen, but from their perspective, the exact opposite has occurred: everyone else in the meeting has frozen, and they remain the same. The world moves on without us when we have become disconnected, while our own frame of reference remains frozen in space and time—a *tableau mort* of failed efficiency.

The norming of user practice around a promise of error-free communication and unimpeded connectivity obscures from view an important element in our everyday encounters with these efficiency tools, and that is: missteps, awkward moments, and communicative glitches provide a very common and very real context for what it means to be "on Zoom." Drawing upon Stein's (2017) discussion of the "lapse" between promised and lived experience within technological regimes, Della Ratta (2021) calls attention to "the discrepancy between the promise of seamless real-time delivery, efficiency, and speed that comes embedded with assumptions of the digital, and the much more messed up and messy reality made of glitches, interrupted sound bites, frozen frames, and abrupt disconnections." Shame, Della Ratta argues, figures prominently in the lived experience of these discrepancies. Here is yet another mapping of the uneasy zone of contact between embodied realities and the informatic efficiencies demanded by the platform itself, not unlike those early pandemic cautionary tales of users who failed to keep their bodily functions in check, out of frame, and off-mic. But the awkwardness of this encounter, Della Ratta argues, need not fold the user back into a disciplinary regime of platform logic; these awkward, disorienting moments can, in fact, mark a zone of resistance to the logic of maximum performance—a point of friction that we should attend to, rather than shame or ignore. Failure (Halberstam 2011) and disorientation (Ahmed 2006) can and do create openings to otherwise excluded practices and encounters. Employing a metaphor of inoculation appropriate to life in those early days of Covid-19, Della Ratta writes:

> Why don't we take the shame element out of the picture (so to speak), and embrace the awkwardness as a vaccine against hyper-performativity and self-burnout? . . . Why not make a space for potential flourishing out from our own self-acknowledged impotence and collective discomfort? How about we just make the best, make maximum awkwardness, make shameless fun, of this one massive "awkward moment" that we're living in?

Such encounters with and through Zoom may indeed create an opening for a mode of connection that does not reduce to connectivity, however awkward, tentative, or unstable that moment might be.

Meeting Disconnected

While the protocological demands of digital interfaces come into play in this discussion, it is likewise important to highlight that our calling into question the frame that allowed Zoom and other videoconferencing platforms to dominate our social, personal, educational, and economic interactions during the pandemic does not reduce to a matter of "good" or "bad" platform architecture or interface design. Rather, the frame itself is indicative of the logic that adapts neoliberalism to the demands of data capitalism. The imperative to perform, to adapt seamlessly, is rooted in the first principle that capital must be *productive*. As Lovink (2022) notes, "There is an imperative here, and it's about productivity and efficiency, not software. As one article quipped 'you don't hate Zoom, you hate capitalism'" (42).[4] It did not matter that in the spring of 2020, we were suddenly thrust into an unstable social world with uncertainty and fear, managing our families and jobs from our living spaces. For those of us whose labor already consisted primarily of adding to or managing information flows, we remained *peculiar commodities* (Marx 1992), worth only what we can produce, even when the barriers to production are deeply strained. And we should not forget the additional emotional labor placed on teachers and faculty, particularly at the start of the pandemic, who not only struggled to move their students online but also strained to keep some semblance of normalcy in play within the disrupted lives of their students, all the while trying to navigate the demands of new tools and technologies. What that meant for many faculty was, after that rapid shift online in March 2020, numerous formal and informal in-services and workshops to learn how to design online courses that would keep students engaged and actively learning. This labor was felt even more heavily by the most precarious within academia, such as contract and adjunct faculty.[5] Solutions to a crisis in course delivery and an epidemic of disengaged students came in the form of technological tools that were both dependent on the labor of instructors to institute them and the assuredness that such tools could become normalized, regardless of their impact or the invasiveness inherent in their design. Indeed, even when the world shut down and millions of workers were sent to labor from home (or some other "remote" location), the expectations of performance and productivity rooted in the culture of capitalism followed closely behind. Cameras and video calls moved from serving as an amusing novelty to operating as a means of surveillance, as workplaces began requiring cameras to be on at all times during working hours, regardless of the messy demands of home and family life. Students at many institutions faced similar constraints with professors demanding that cameras stay on or the forced use of monitoring software to proctor exams. Our performances became those of productive workers under a watchful digital eye, in spite of the mental and physical strain that came with trying to keep up with the show.

Under the regime of videoconferencing, surveillance of labor takes the form of what Søren Pold refers to as the *Zoomopticon*:

> The condition in which you cannot see if somebody or something is watching you, but it might be the case you're being watched by both people and corporate software. Zoomopticon has taken over our meetings, teaching and institutions with a surveillance capitalistic business model without users being able to define precisely how this is being done. (quoted in Lovink 2022, 43)

Under such conditions, it is not surprising that workers and students found ways to challenge these rules, with many developing hacks to trick monitoring software. Lovink (2022), for example, points to interventions that turned the logic of maximum performance against itself, such as digital artist Sam Lavigne's (2021b) *Zoomescaper*, an audio driver that allows users to "self-sabotage your audio stream, making your presence unbearable to others" with effects including "Echo," "Bad Connection," "Upset Baby," and "Urination" (47). Lavigne's project, along with his earlier Zoom Deleter (2021a) program, which "continually checks for the presence of Zoom on your computer, and if found, immediately deletes it," operates within the same code logic that supports communicative capitalism but, in place of smooth flows of information, offers tools of blockage and obfuscation. Likewise, the simple act of turning off one's camera, while itself an implementation of a user control, functioned as a mild, yet effective Luddite refusal, a small but meaningful counter-hegemonic strategy for resisting the demands of a capitalist performance of productivity. At the same time, the line between escape and capture is hardly clear—"Zoom bots," at one point designed for spam attacks, or to spoof class attendance for truant students, now are supported functions within Zoom's API as yet another efficiency tool ("Meeting Bots" n.d.)

For teachers and faculty, institutions offered up Zoom as a solution to the pandemic but also framed the tool as a means of providing a more engaging learning experience. To the contrary, however, regardless of their efforts, many faculty found that students had at their fingertips the tools to *disengage* from the same institutional networks that reached out into their homes. As Weinburgh

(2022) notes in her case study of student teachers during the pandemic, from a teacher's perspective, students' lack of engagement, regardless of the effort put in by the instructor, "felt like failure"; as one student teacher commented: "*Field of Dreams*—not what happened for me. I built it but they did not come." Weinburgh's case study is far from unique. During the spring and fall semesters of 2020, many students reported feeling less engaged in Zoom classes compared to their face-to-face classes and less satisfied with their ability to connect with faculty or fellow students (Serhan 2020). Discussions of engagement in online learning frequently take the form of a post-mortem on student *disengagement*, exploring the means by which teachers and faculty can make better use of the medium and its affordances to increase engagement. Such was certainly the case during the Covid-19 pandemic, as educational researchers attempted to diagnose and cure the cause of students' failure to perform as engaged learners (Maimaiti et al. 2021; Rolins et al. 2022; Martin and Borup 2022). Maimaiti et al. (2021), for example, discuss the importance of "online social presence" (among other factors) in sustaining engagement. The camera, they assert, offers an important means for creating that sense of presence, thus they advise faculty: "to remind their students to switch on their cameras before each session. Webcams also show a person's face to other people on the video call, which can help *less disciplined students* to stay focused in the videoconferencing class" (16, emphasis added). While offering suggestions to mitigate the potentially negative impact of cameras-on, such as using a Zoom background to hide "undesirable background objects" or allowing for occasional cameras-off participation to accommodate a student's self-consciousness when speaking, it is clear from their analysis that in the videoconferencing classroom, the camera is itself a tool that simultaneously measures and monitors the engagement it seeks to create (Maimaiti et al. 2021, 16). In effect, the webcam is a disciplinary tool, one that enforces its own terms for what it means to be (digitally) engaged.

Not that we are surprised by these conclusions. As Kuntsman and Miyake (2022) note, once "student engagement" moves to an online environment, its varied forms collapse into measures that are compatible with and synonymous for *digital* engagement. But what of the student who insists on maintaining camera-off during Zoom sessions? Certainly this phenomenon became increasingly common as the familiarity with the interface increased, leaving instructors with one of two options: either search for more draconian measures to ensure that students kept cameras on (e.g., lack of an attendance grade for failing to turn on one's camera) or accept the fact that they would be instructing to a grid of black boxes. Note that in Maimaiti et al.'s (2021) discussion of allowing cameras-off for the shy, self-conscious, or anxious student, their allowances parallel modes of *disability accommodation* for a student not otherwise able to perform to standard (as opposed to, for example, the "less disciplined" student). Shyness and introversion—or for that matter, a desire *not to be seen* by a public eye in one's private settings—becomes a form of pathology in an increasingly networked society: a "deviation from the perceived norms of connectivity" (Stäheli 2021, 5). Other modes of stigma attach themselves to the act of refusing to engage with the camera—an admission once again to the affective weight of shame that measures one's fall from the norm of the controlling and controlled entrepreneurial subject. As Della Ratta (2021) notes, having one's camera *on* communicates one's access to an excess of social capital: "a luxury option that under-privileged people, migrants, proletarians, or anyone experiencing a physical or mental disorder (including stress and anxiety), simply cannot afford. Rather than helping class participation or facilitating the learning experience, it has become an exclusive, divisive, and ultimately oppressive, tool." Always *on*, always connected, presents itself as a norm—and it is precisely this normalization of connectivity that has motivated an emerging field of "disconnection studies" (e.g., Light 2014; Karppi et al. 2021; Kuntsman and Miyake 2022) to explore and challenge those assumptions and the ideologies that drive them. Here, something akin to an ableist erasure of nonconforming bodies expresses itself in the assumption that to avoid the camera's gaze is a form of pathology, and to resist its penetrating stare into a domestic space is a symptom of a falling away from a normalized middle-class home, free from "undesirable background objects."

So, what is our caring and compassionate faculty member to do, then, as they attempt to accommodate a class that has expressed a desire to remain cameras-off? A strange inversion of the platform's control logic has occurred, turning the tools of efficient communication into a means of errant refusal. With a click of an icon, the student makes use of the affordances of connectivity to engage in a strategy of disconnection. This moment is what Stäheli (2021) refers to as a practice of *undoing networks*: "neither a simple negation of networks, nor just unplugging from networks, but practices within networks that question their ethico-political impositions of connectivity" (6). Cameras-off, then, functions less as a measure of student disengagement that a faculty must remedy but rather as a reminder of the unstable connections that a norm of "always on" connectivity both creates and attempts to erase. Faculty may feel frustrated, they may bemoan their students' lack of engagement—they may even offer compassion for the disruptive circumstances in which they find themselves as students forced into online education through a global pandemic. But the platform logic that normalizes (digital) engagement as a good remains for the most part unchallenged. In these moments of students "using networks to disconnect from networks," however, that challenge comes to the fore (Stäheli 2021, 14).

Incorrect Meeting Passcode: Please Try Again

While "cameras off" made visible the instability of connection that the platform both creates and seeks to erase, Zoom also offered an arena for practices that revealed connectivity in its most virulent form. Literally days after the WHO declared Covid-19 a global pandemic and millions of individuals found themselves on Zoom for work, school, and social events, reports began to emerge of a growing trend of "zoombombing," in which uninvited participants would begin to screen share inappropriate content, often racist, sexist, and/or pornographic, with the intent of disrupting meetings and causing harm. As one interviewed subject notes in a *New York Times* article on zoombombing from March, 2020: "It makes us all feel pretty helpless in an already unstable time. . . . If I'm going to be asked to live in Zoom University or Zoom Tavern, then I want to know that it's secure for everyone" (Lorenz 2020). As Nakamura et al. (2021) note, zoombombing is not "mere" trolling but rather it operates as a form of "memetic warfare" that hides in plain sight its racist, sexist, and homophobic violence behind a "for the lulz" history of (equally racist, sexist, and homophobic) disruptive online pranking (34). More often than not, the institutional response to zoombombing—both from Zoom as a corporate entity and from the organizations and institutions that often host these events—has taken the form of improved modes of user control over meetings (host functions and passcodes, for example): a neoliberal responsibilization of practices intended to eliminate the "noise" of unwanted participants, rather than an attempt to address the systemic discrimination and social inequity that networked platforms can and do perpetuate (Ali 2021; Nakamura et al. 2021).

In this regard, zoombombing highlights the unstable connections that are both inherent in and obscured by the short-term embrace of technology as a solution to the pandemic as well as the longer-term ideologies that inform a logic of Zoom. First and foremost: those who are targeted for zoombombing attacks are predominantly those groups who already maintain more precarious relationships to technology (Elmer et al. 2021; Nakamura et al. 2021; Lee 2022). In an analysis of YouTube compilation videos of zoombombing attacks, Elmer et al. (2021) found that 86 percent of their sample contained "racist, misogynist, homophobic, and other toxic content." In addition to classroom settings, which bore the largest brunt of these attacks, communities most in need of mediated support during times of social distancing found their networks most likely targeted for the "affective violence" of zoombombing, from twelve-step recovery meetings to Black scholar peer groups (Elmer et al. 2021; Nakamura et al. 2021).

Elmer et al.'s (2021) quantitative research likewise confirms Nakamura, Stiverson, and Lindsey's (2021) assertion that the response to zoombombing has tended to erase this targeting of precarious and marginalized populations, focusing instead on issues of control. They note that in Twitter discussions of zoombombing that emerged in the midst of that early wave of attacks, "racism and abuse was never addressed as a harm by Twitter users who advised on how to mitigate the impact of zoombombing. In short, we found that the abusive practice of zoombombing was largely framed as disruptions of meetings, education, or work, enabled by technological insecurities" (Elmer et al. 2021). By maintaining a focus on restoring the efficiency and productivity that a Zoom logic asserts, those whose privilege either removed them from the crosshairs of these assaults or inoculated them against the real violence of these attacks became (wittingly or unwittingly) complicit in maintaining connectivity at the expense of obscuring its uneven distribution. Finally, we note an important intersection between marginalized groups and academic faculty in the wake of the pandemic and zoombombing—folks who most certainly took on additional emotional labor in managing and supporting their students while also serving as targets for zoombombers. Further, considering the instability of professional, educational, and social networks, victims of zoombombing in the academic setting faced their own trauma while managing the trauma of their likewise marginalized students. In some ways, then, zoombombing reinforced disconnection from their more privileged counterparts on campus.

We would note as well the confluence of factors that links the instability of networks to the same logic that asserts hyperconnectivity as a normative good. While Zoom provided the platform upon which to launch these attacks, they could only occur within the complex, hyperconnected media ecology that included online sites like Reddit, Twitter, and 4chan. One was just about equally likely to encounter zoombombing Tweets that were coordinating attacks as one was to encounter "helpful" defensive advice, with the results skewing far higher on Reddit (Elmer et al. 2021). In short, while zoombombing attempted to exploit weaknesses in the security of meetings to target vulnerable groups with acts of hatred and violence, such errant practices were only possible by way of the hyperconnective networks that are indeed inescapable and embedded in every aspect of our lives. While the frame of Zoom forces such transgressive acts out of the picture, so to speak, as a literal violation of their terms of service and terms of use, such potential is far from errant and far from marginal; it is, rather, another symptom of the operative instability that makes "always on" connectivity possible.

Conclusion: Let Me Log Out and Then Come Back In Again

As unstable, unpredictable, and inequitable as these networks of connectivity have proven to be, we have also found in our working with—and against—the logic of Zoom hopeful moments of possibility, camaraderie, and meaningful connection. And that should come as no surprise to those of you who, like all of us who have contributed to this volume, have come to know Zoom *by way of Zoom*, creating "affective encounters" with the platform that likewise informed our ways of understanding daily life on a videoconferencing platform (Bucher 2018, 94). And here we may find yet another reason for the strong draw toward autoethnography in this collection—and in many accounts of the Covid-19 pandemic (Herrmann and Adams 2020): we, as scholars and academics, were all quite literally caught up in the daily practice of being "on Zoom" throughout the pandemic and experienced firsthand the very things this volume attempts to put in words. Our approach in this volume, much like Olli Pyyhtinen's (2016) expansion of C. Wright Mills's (2000) classic theory of the sociological imagination, has been to consider the more nuanced, complicated, and deeply interconnected pathways and crosswires that make up social experience—online, in person, and entangled in-between. Mills argued for an understanding of the intersection of history and biography, moving from the micro experiences to reflect on the macro events that shape them. Pyyhtinen suggests we don't simply move from micro to macro, but rather that we traverse a complicated, rhizomatic web of interactions that makes up our social experiences—how we relate to, interact with, and are shaped by not just fellow social actors but to all matter, human or otherwise. To ignore our own personal experiences on Zoom—from the first stay-at-home orders to the present moment—would be to flatten out the complexity of the pathways of knowing that have informed our shared understanding of the *place* of Zoom in everyday life over the past three years.

For some, the pandemic created an opportunity to develop for the first time sustained, online relationships and to engage with communities in digital landscapes. For others, having online forums as a place for personal and professional relationships was familiar terrain, but now the rapid expansion of videoconferencing platforms added additional options and tools for relationship building. Each of us (Mark and Cassandra) came onto Zoom in the spring of 2020 with some previous experience with online communities. For one of us, it was the first time they had taken part in regular, online community building and networking specifically related to their professional life beyond casual conversations on spaces like Twitter. For the other, it was a return to their early days on the internet, when they were active on list-servs and in real-time, text-based "virtual communities" in an attempt to explore novel opportunities for scholarly collaboration, only now those efforts were supported by videoconferencing tools. Without question, though, Zoom—and academic life during pandemic times—brought us together in ways that likely would not have emerged otherwise.

Our professional relationship began in person at the Cultural Studies Association (CSA) Annual Conference in New Orleans the summer before Covid-19. That next year, as with many professional organizations, the conference moved online and used Zoom as its platform. We all presented our work in the little windows and grid layouts that became such a common venue for professional and personal encounters in the months and years to follow. As our own professional lives became shaped by the logic and institutional pressures that pushed our work onto Zoom, it seemed "only natural" to make use of these tools to support the collaborative efforts of the CSA New Media and Digital Cultures working group—a structure that, once again, predated the pandemic but came to be something more than it had ever been once so many of us moved online. We began to meet as a group via Zoom regularly—every four to six weeks during that early pandemic period. Though many of us were somewhat familiar with each other from previous conferences, meeting regularly over Zoom built up strong professional connections as well as affective bonds, as we all lamented on the state of the world while bumbling along learning how to mute and unmute, make snarky comments in the chat, and share screens with timely, relatable, and sarcastic memes. Connecting regularly through Zoom allowed for, we argue, two important developments:

1. The connection between folks of like interests to move those shared interests into shared projects (themed issues in *M/C Journal* and the CSA journal *Lateral*; an ongoing podcast series; and, of course, this volume itself).

2. The development of meaningful formal and informal mentorship supports for graduate students. (The working group developed a formal mentorship program during this same period, pairing grad students with mentors based on interests and academic/professional needs for a year long relationship. This program is still ongoing).

These regular meetings allowed space for laughter, frustration, and cheering each other on. New projects were regularly pitched, workshopped, and planned by any and all who had an inkling of interest, regardless of level of education or experience; if anything, more senior members of the group made regular space for the graduate students to show an interest—and unmute themselves. Though we cannot speak for the other members of the group, it was the culmination of these opportunities presented through those two developments that led to the creation of this current collection.

In our final days of developing this volume of work, we reflected on just how easily the working group became a part of our regular lives on Zoom, but more specifically, how it became a community that supported the members' needs. Indeed, the group itself had not been very active prior to the pandemic; it wasn't until we were all relegated to our remote, digital spaces that we truly began to build a real community—one that continues to this day and the members of whom we thank again for their ongoing support in this work. Much like Cochrane and Kenny's concept of "networked togetherness" in this volume, the working group became a collegial and spirited group of academic orphans who found a home—and many a project—as we moved past professional performances to genuine connections.

For us as editors, we found our own working groove over Zoom; having only met once in person, we met regularly online over the course of two years to develop and edit our *M/C Journal* special issue and now this volume. Our professional relationship expanded from one of mentor/mentee to friendship, as we learned new things about the other's life through a variety of hilarious, and sometimes serious, Zoom encounters that often led to the sharing of our personal circumstances that were affecting our workspace, our home life, or our ability to "perform" professionally at all. Working together to accommodate family emergencies, precarious work environments, or sudden life changes—even through our screens—created a different Zoom experience for us than we had been facing in much of the rest of our professional—and personal—lives. We could often tell how things were going for the other person simply based on which room or office they were Zooming from that day. In a hilarious series of events mere weeks before this manuscript was due to be submitted, we could not manage to get on Zoom together for our weekly Friday meeting. From time zone changes to wrong Zoom calls to internet connections that wouldn't work while fumbling to get online at an in-laws' house, our connection suddenly seemed less than stable and at a critical professional juncture. And yet, the following week, we connected as always, with much progress to share. Our connection had moved beyond the digital realm, even if our meetings were limited to it. Perhaps most importantly, our connection also challenged the ableist demands of capital to be productive—and perfect—regardless of circumstance. We encouraged each other to take breaks, to rest, and we asked for time when we needed it. We supported each other when facing big events, and we made space for quiet moments of sadness when those also, unfortunately, arose. Indeed, in spite of the global push for a "return to normal," which seemed to include a rejection of adapted productivity environments, a work/life balance, and a slowing down to stay rested and safe, we challenged those conventions in our own methods of writing, researching, and developing this volume. Our own "networked togetherness" prioritized collaborative and supportive work over traditional academic environments that all too often focus on perfectionism, hyperproductivity, and inequitable workloads—particularly when it comes to junior and senior academics cowriting and editing.

Throughout this volume, and this chapter, our writing has gestured toward a "we"—Mark and Cassandra—but we also see ourselves in the experiences of the other contributors to this collection, as well as the countless academics and writers who adapted as best they could to the isolation and the continued professional pressures (strange bedfellows as they were) of working and living on Zoom. Indeed, we all did what we could to cobble together some kind of life on Zoom—something many of us are still trying to navigate. Much of the "we" that echoes throughout this chapter is our call of solidarity into the dark screen void of endless Zoom meetings, classes, and team networking events.

Our experiences are not the same for everyone, but as has been demonstrated in this collection, we—and you—are certainly not alone.

On one hand, we—Mark and Cassandra—might say, somewhat cynically, that as heavily institutionalized subjects of the neoliberal university, we had answered the call of productivity, lifting ourselves up by our digital bootstraps to get back to the good work of performing efficiently within a network logic of hyperconnected, laboring subjects. But something else happened as well, and reducing the narrative of our experience on and through Zoom—and in the midst of a global health crisis—to a critical assessment of a logic of maximum performance does not ring true to our "affective encounter" with the connections we made, the co-labor we performed, and the experience of building what we built—in all of its tenuous, unstable glory. We found ourselves forming networked relationships and deploying online resources (often in unanticipated and perhaps unsanctioned ways) to support the immediate needs of an ad hoc and loosely assembled community of teachers, scholars, students, and academics. The contingency of the connections we made through the working group (and beyond) kept us focused on how best to support our online community members, and ourselves. Our working group, our editorial meetings, our basic check-ins: these were our responses to an overarching systemic failure in the social systems that were meant to sustain us. The up-front acknowledgment of our overtaxed, overconnected, and overworked selves, the vulnerability that came from working and being together through Zoom: all of that tenor and flavor cannot be ignored.

And so, we do not return to normal; we finish this piece, this volume, this work, changed as scholars, yes, but also people, and reminded of the unstable connections that can, if supported, adapted, and held carefully and meaningfully, keep us grounded.

Notes

1 We acknowledge the ableist implications of a term such as "blind spot," yet given the *video* in videoconferencing, and the epistemological primacy of the visual it assumes, we use this term with both intention and care. We would also note a similar ableist bias in the common usage of "muting" for silencing of mics, which we employ with equal caution throughout this chapter.

2 For further discussion of the ideological and technological importance of error handling within cybernetic control systems, and how error itself can mark a moment of aesthetic and political potential, see Nunes (2012).

3   It is worth noting in this context that even proficient users have a fairly limited understanding of what their mute button does and the degree to which platforms still monitor microphone data even when muted (Yang et al. 2022). Yang et al. (2022) remark that even though this data sampling may fall in line with the intended uses of the platform (e.g., Zoom must sample microphone input if it is going to provide that helpful reminder "You are muted. Press Alt+A to unmute"), this gap between user expectations and data realities is problematic and may indeed suggest legitimate privacy concerns.

4   While Lovink (2022) does not provide a direct citation for this quote, nor were we able to find it in our searches online, it is, of course, a play on the oft-memed quote frequently attributed to (again, without citation) Slavoj Žižek: "You don't hate Mondays, you hate capitalism."

5   While the pandemic has forced an increase in mental health supports on many university campuses for students, there continues to be a gap in support for faculty, and in particular, coverage for temporary/adjunct faculty, or a more intentional consideration of the emotional toll the pandemic took on educators overall.

References

Ahmed, Sara. 2006. *Queer Phenomenology: Orientations, Objects, Others* Durham: Duke University Press.

Ali, Kawsar. 2021. "Zoom-ing in on White Supremacy : Zoom-Bombing Anti-Racism Efforts." *M/C Journal* 24, no. 3. https://doi.org/10.5204/mcj.2786.

Bucher, Taina. 2018. *If... Then: Algorithmic Power and Politics*. New York: Oxford University Press.

Butler, Judith. 2016. *Frames of War: When is Life Grievable?* New York: Verso.

Della Ratta, Donatella. 2021. "Teaching Into the Void: Reflections on 'Blended' Learning and Other Digital Amenities." *INC Longform*, January 6, 2021. https://networkcultures.org/longform/2021/01/06/teaching-into-the-void/.

Elmer, Greg, Stephen J. Neville, Anthony Burton, and Sabrina Ward-Kimola. 2021. "Zoombombing During a Global Pandemic." *Social Media + Society* 7, no. 3. https://doi.org/10.1177/20563051211035356.

Halberstam, Jack. 2011. *The Queer Art of Failure*. Durham: Duke University Press.

Herrmann, Andrew and Tony Adams. 2020. "Living Through/With COVID-19: Using Autoethnography to Research the Pandemic." *UC Press Blog*. September 1, 2020. https://www.ucpress.edu/blog/51794/living-through-with-covid-19-using-autoethnography-to-research-the-pandemic/.

Karppi, Tero, Urs Stäheli, Clara Wieghorst, and Lea P. Zierott. 2021. *Undoing Networks*. Minneapolis: University of Minnesota Press.

Kuntsman, Adi and Esperanza Miyake. 2022. *Paradoxes of Digital Disengagement: In Search of the Opt-Out Button*. London: University of Westminster Press.

Lavigne, Sam. 2021a. "Zoom Deleter." https://lav.io/projects/zoom-deleter/.

Lavigne, Sam. 2021b. "Zoomescaper." https://lav.io/projects/zoom-escaper/.

Lee, Claire Seungeun. 2022. "Analyzing Zoombombing as a New Communication Tool of Cyberhate in the COVID-19 Era." *Online Information Review* 46, no. 1: 147–163. https://doi.org/10.1108/OIR-05-2020-0203.

Light, Ben. 2014. *Disconnecting with Social Networking Sites*. New York: Palgrave.

Lorenz, Taylor. 2020. "'Zoombombing': When Video Conferences Go Wrong." *New York Times*, March 20, 2020. https://www.nytimes.com/2020/03/20/style/zoombombing -zoom-trolling.html.

Lovink, Geert. 2022. *Stuck on the Platform: Reclaiming the Internet*. Amsterdam: Valiz.

Maimaiti, Gulipari, Chengyuan Jia, and Khe Foon Hew. 2021. "Student Disengagement in Web-Based Videoconferencing Supported Online Learning: An Activity Theory Perspective. *Interactive Learning Environments* 31, no. 8: 4883–4902. https://doi.org /10.1080/10494820.2021.1984949.

Martin, Florence and Jered Borup. 2022. "Online Learner Engagement: Conceptual Definitions, Research Themes, and Supportive Practices." *Educational Psychologist* 57, no. 3: 162–177. https://doi.org/10.1080/00461520.2022.2089147.

Marx, Karl. 1992. *Capital: A Critique of Political Economy*, Volume 1. New York: Penguin.

"Meeting Bots: Accessing Media Streams." n.d. *Zoom Developers*. https://developers .zoom.us/docs/zoom-apps/guides/meeting-bots-sdk-media-streams/ (accessed November 18, 2023).

Mills, C. Wright. 2000. *The Sociological Imagination*. 40th Anniversary ed. Oxford: Oxford University Press.

Nakamura, Lisa, Hanah Stiverson, and Kyle Lindsey. 2021. *Racist Zoombombing*. New York: Routledge.

Nunes, Mark. 2012. "Error, Noise, and Potential: The Outside of Purpose." In *Error: Glitch, Noise, and Jam in New Media Cultures*, edited by Mark Nunes, 3–24. New York: Continuum.

Pyyhtinen, Olli. 2016. *More-Than-Human Sociology: A New Sociological Imagination*. London: Palgrave Macmillan.

Rolins, Dilian, Lea Herbert, and Galaxina Wright. 2022. "Logged In, Zoomed Out: Creating & Maintaining Virtual Engagement for Counselor Education Students." *Journal of Technology in Counselor Education and Supervision* 2, no. 1. https://doi.org /10.22371/tces/0016.

Serhan, Derar. 2020. "Transitioning from Face-to-Face to Remote Learning: Students' Attitudes and Perceptions of Using Zoom during COVID-19 Pandemic." *International Journal of Technology in Education and Science* 4, no. 4: 335–342. https://eric.ed.gov/?id=EJ1271211.

Stäheli, Urs. 2021. "Undoing Networks." In *Undoing Networks*, edited by Tero Karppi, Urs Stäheli, Clara Wieghorst, and Lea P. Zierott, 1–30. Minneapolis: University of Minnesota Press.

Stein, Rebecca. 2017. "GoPro Occupation: Networked Cameras, Israeli Military Rule, and the Digital Promise." *Current Anthropology* 58, no. S15. https://doi.org/10.1086/688869.

Weinburgh, Molly H. 2022. "'Students Were Just Sticky Notes on JamBoard': A First-Year Biology Teacher's Story of 2020–2021." *School Science & Mathematics* 122, no. 5: 235–246. https://doi.org/10.1111/ssm.12539.

Yang, Yucheng, Jack West, George K. Thiruvathukal, Neil Klingensmith, and Kassem Fawaz. 2022. "Are You Really Muted? A Privacy Analysis of Mute Buttons in Video Conferencing Apps." arXiv:2204.06128. April 13, 2022. https://doi.org/10.48550/arXiv.2204.06128.

Contributors

Rachel Buchanan is an associate professor in education at the University of Newcastle, Australia. Her research interest centers on social justice and equity in education, leading to her scholarship on educational technologies, educational policy, digital identity, and the equity implications of the increased use of digital technologies for education and career access. https://orcid.org/0000-0003-3594 -1110.

Earvin Charles B. Cabalquinto is an Australian Research Council DECRA Research Fellow and Senior Lecturer at Monash University. He is the author of *(Im)mobile Homes: Family Life at a Distance in the Age of Mobile Media* (2022). His expertise lies in the intersecting fields of digital media, mobilities, migration, and aging research. His current projects explore the production of the digital divide in an increasingly transnational and networked era.

Heather J. Carmack is a senior associate consultant in the Robert D. and Patricia E. Kern Center for the Science of Health Care Delivery at the Mayo Clinic. Her research centers on communication about patient safety and identifying ways to help clinicians become more competent communicators. She also studies the role of organizational processes, policies, and practices in workplace communication.

Alexis-Carlota Cochrane is a PhD candidate in the Department of Communication Studies and Media Arts at McMaster University in Ohròn:wakon (Hamilton, Ontario). Alexis (she/they) researches at the intersection of identity, technology, and digital culture. Her latest cowritten essays are entitled "Discourses on Cybersecurity. The Politics of the Data Breach as a Security Crisis" (2022) and "Constellations of Community, Care, and Knowledge: A Collection of Vignettes from Pandemic Times" (2023). She can be found online at @alexis_carlota.

Ricky D'Andrea Crano is a research specialist in the Humanities Center and a lecturer in Film and Media Studies and Anthropology at the University of

California, Irvine. His research and teaching span a range of topics across the politics and aesthetics of digital media and algorithmic culture. His book-in-progress is titled *Affect and Account: Liberal Discourse, Fascist Desire, and the Production of the Algorithmic Unconscious.*

Rory Davis is a secondary creative arts teacher for the NSW Department of Education, specializing in visual arts, drama, and photography. He is a PhD candidate at the University of Newcastle in the College of Human and Social Futures, currently researching soft skills and visual arts curriculum in Australian secondary schools.

Jocelyn M. DeGroot is a professor in the Department of Applied Communication Studies at Southern Illinois University Edwardsville. Her research interests include computer-mediated communication and communicative issues of death and dying. She also examines dynamic tensions in interpersonal relationships such as MIL/DIL relationships and motherhood.

Craig Fahner is a visiting assistant professor in the Integrated Design and Media program at New York University. His research and creative work questions and reimagines the ways that media technologies shape everyday life. His work has been shown in various venues internationally, including the Alberta Biennial of Contemporary Art and the Device Art Triennial in Zagreb. He is a co-investigator in the Data Fluencies project, a Mellon Foundation-funded research initiative that is working to establish critical public literacies around the impacts of data-driven technologies.

Kathryn Grushka is a researcher and senior lecturer in education at the University of Newcastle, New South Wales, Australia. Kathryn is a fiber artist/painter and academic. She researches in the areas of visual art and design education; artmaking and subjectivity; arts/health and arts/science; and digital visual learning. Her inquiry draws on narrative and arts-based methods; the performative work of image construction; critical and social inquiry; and visual performative pedagogies as they surround the contemporary subject. Her scholarship is grounded in neuroscience, enactive cognition, Deleuzoguattarian thinking, and new materialism. She publishes nationally and internationally and works with a range of research and editorial teams. https://orcid.org/0000-0002-4228-3606; Instagram: @kath.grushka.

Jacquelyne Thoni Howard is a professor of practice at the Connolly Alexander Institute for Data Science at Tulane University. She teaches courses on data literacy and technology studies. Her research focuses on digital humanities, history of borderlands and families, and history of surveillance and data. She is a cofounding editor of the "Feminist Pedagogy for Teaching Online" guide (https://feminists-teach-online.tulane.edu/). Read more at @thonihoward | https://thonihoward.wp.tulane.edu/ | https://orcid.org/0000-0003-3574-3405.

Theresa N. Kenney is a PhD candidate and teaching fellow in the Department of English and Cultural Studies at McMaster University in Ohròn:wakon (Hamilton, Ontario). Theresa's (she/her) research explores asexual, aromantic, and platonic relationalities in queer Asian North American cultural production. Her latest cowritten essay is entitled "Constellations of Community, Care, and Knowledge: A Collection of Vignettes from Pandemic Times" (2023). She can be found online at @ToPoliticise.

John A. McArthur is professor and chair of the Department of Communication Studies at Furman University and the author of *Digital Proxemics: How Technology Shapes the Ways We Move* (2016). At Furman, Dr. McArthur teaches courses in social and mobile media and the impacts of digital technologies on our lived experience. His research focuses on proxemics (the use of space) and information design with a particular interest in the ways that digital technology influences our interactions with and in gathering spaces.

Mark Nunes is professor of interdisciplinary studies at Appalachian State University. He is author of *Cyberspaces of Everyday Life* (2006) and editor of a collection of essays entitled *Error: Glitch, Noise, and Jam in New Media Cultures* (2012). He has written extensively at the intersection of cultural studies and new media studies, with recent work exploring the co-constitution of human and algorithmic agencies.

Daniel Paul O'Brien is a lecturer in film and digital Media at the University of Essex. His interests and areas of research span across film, video game studies, interactive media art, video essaying, and AI. He has published work in all these subjects. Recent publications include "Digital Love: Through the Screen/Of the Screen" (2023), "Media Ownership and Digital Authenticity in Slum TV" (2022), and "Extant's Flatland: Disability and Postphenomenological Narrative"

(2020). He is currently working on a monograph titled *Postphenomenology and Narrative across Cinema, Interactive Art and Gaming.*

Cassandra Ozog is an instructor in the Department of Sociology and Social Studies at the University of Regina, in Treaty 4 Territory (Regina, Canada). She recently completed her interdisciplinary PhD in sociology and media studies. Her research interests include representations of mental illness and disability in the media, digital methodologies, and popular culture, specifically horror and sci-fi film, fiction, and television.

Evgeniya Pyatovskaya is a PhD student at the University of South Florida. Her research interests are in organizational, intercultural communication, and feminist perspectives on communication and organizational leadership and resilience. Evgeniya is a Fulbright FLTA and Atlas Corps alumna with an extensive international work experience that she builds on in her scholarly work. One of her most recent projects looks at feminist narratives of resilience in the context of the pandemic and how culture-specific notions of agency influence what resilience looks like and what its long-term consequences can be.

Susan A. Sci is associate professor and chair of the Communication Department at Regis University. Dr. Sci's research focuses on the affective and behavioral impact digital media and technology has on individuals and society. Her recent work addresses how the transition from analog to digital, mobile technology has impacted students' news consumption and literacy. At Regis, Dr. Sci teaches courses on media literacy, platform studies, public memory, gender studies, and critical media analysis.

Heather M. Stassen is the dean of the College of Arts, Sciences and Education at Daemen University. Dr. Stassen's research interests are at the intersection of public controversy and rhetoric. Specifically, she seeks to unpack public opinion and ideology through analysis of contemporary controversies.

Tennley A. Vik is an independent researcher whose work focuses on interpersonal relationships, privacy dilemmas, family relationships, and social support. Her research includes publications in *Personal Relationships, Journal of Family Communication, Sexuality & Culture, Computers and Human Behavior, Communication Studies,* and *Journal of Loss and Trauma.*

Michael Whittington is head of art and design (7-9) and a senior visual arts teacher at an independent boys school in Sydney, Australia. He has taught for over twenty years including in the UK and Australia. He is currently a PhD candidate at the University of Newcastle researching how adolescent boys are developing empathy from their material art practice in a twenty-first century visual arts classroom. -https://orcid.org/0000-0001-9493-2623.

Index